T0353474

Distillation

There are no such things as applied sciences,
only applications of science
Louis Pasteur (11 September 1871)

Dedicated to my wife, Anne, without whose unwavering support, none of this
would have been possible.

Industrial Equipment for Chemical Engineering Set

coordinated by
Jean-Paul Duroudier

Distillation

Jean-Paul Duroudier

First published 2016 in Great Britain and the United States by ISTE Press Ltd and Elsevier Ltd

ISTE Press Ltd
27-37 St George's Road
London SW19 4EU
UK

www.iste.co.uk

Elsevier Ltd
The Boulevard, Langford Lane
Kidlington, Oxford, OX5 1GB
UK

www.elsevier.com

Notices

Knowledge and best practice in this field are constantly changing. As new research and experience broaden our understanding, changes in research methods, professional practices, or medical treatment may become necessary.

Practitioners and researchers must always rely on their own experience and knowledge in evaluating and using any information, methods, compounds, or experiments described herein. In using such information or methods they should be mindful of their own safety and the safety of others, including parties for whom they have a professional responsibility.

To the fullest extent of the law, neither the Publisher nor the authors, contributors, or editors, assume any liability for any injury and/or damage to persons or property as a matter of products liability, negligence or otherwise, or from any use or operation of any methods, products, instructions, or ideas contained in the material herein.

For information on all our publications visit our website at http://store.elsevier.com/

British Library Cataloguing-in-Publication Data
A CIP record for this book is available from the British Library
Library of Congress Cataloging in Publication Data
A catalog record for this book is available from the Library of Congress
ISBN 978-1-78548-177-2

Printed and bound in the UK and US

Contents

Preface

The observation is often made that, in creating a chemical installation, the time spent on the recipient where the reaction takes place (the reactor) accounts for no more than 5% of the total time spent on the project. This series of books deals with the remaining 95% (with the exception of oil-fired furnaces).

It is conceivable that humans will never understand all the truths of the world. What is certain, though, is that we can and indeed must understand what we and other humans have done and created, and, in particular, the tools we have designed.

Even two thousand years ago, the saying existed: "faber fit fabricando", which, loosely translated, means: *"c'est en forgeant que l'on devient forgeron"* (a popular French adage: *one becomes a smith by smithing*), or, still more freely translated into English, "practice makes perfect". The "artisan" (faber) of the 21st Century is really the engineer who devises or describes models of thought. It is precisely that which this series of books investigates, the author having long combined industrial practice and reflection about world research.

Scientific and technical research in the 20th century was characterized by a veritable explosion of results. Undeniably, some of the techniques discussed herein date back a very long way (for instance, the mixture of water and ethanol has been being distilled for over a millennium). Today, though, computers are needed to simulate the operation of the atmospheric distillation column of an oil refinery. The laws used may be simple statistical correlations but, sometimes, simple reasoning is enough to account for a phenomenon.

Since our very beginnings on this planet, humans have had to deal with the four primordial "elements" as they were known in the ancient world: earth, water, air and fire (and a fifth: aether). Today, we speak of gases, liquids, minerals and vegetables, and finally energy.

The unit operation expressing the behavior of matter are described in thirteen volumes.

It would be pointless, as popular wisdom has it, to try to "reinvent the wheel" – i.e. go through prior results. Indeed, we well know that all human reflection is based on memory, and it has been said for centuries that every generation is standing on the shoulders of the previous one.

Therefore, exploiting numerous references taken from all over the world, this series of books describes the operation, the advantages, the drawbacks and, especially, the choices needing to be made for the various pieces of equipment used in tens of elementary operations in industry. It presents simple calculations but also sophisticated logics which will help businesses avoid lengthy and costly testing and trial-and-error.

Herein, readers will find the methods needed for the understanding the machinery, even if, sometimes, we must not shy away from complicated calculations. Fortunately, engineers are trained in computer science, and highly-accurate machines are available on the market, which enables the operator or designer to, themselves, build the programs they need. Indeed, we have to be careful in using commercial programs with obscure internal logic which are not necessarily well suited to the problem at hand.

The copies of all the publications used in this book were provided by the *Institut National d'Information Scientifique et Technique* at Vandœuvre-lès-Nancy.

The books published in France can be consulted at the *Bibliothèque Nationale de France*; those from elsewhere are available at the British Library in London.

In the in-chapter bibliographies, the name of the author is specified so as to give each researcher his/her due. By consulting these works, readers may gain more in-depth knowledge about each subject if he/she so desires. In a reflection of today's multilingual world, the references to which this series points are in German, French and English.

The problems of optimization of costs have not been touched upon. However, when armed with a good knowledge of the devices' operating parameters, there is no problem with using the method of steepest descent so as to minimize the sum of the investment and operating expenditure.

1

Theoretical Plates in Distillation, Absorption and Stripping – Choice of Type of Column

1.1. General

1.1.1. *Definitions*

A theoretical plate is characterized by the fact that the vapor and the liquid it leaves behind are at equilibrium in terms of pressure, temperature and composition. Each theoretical plate is a point on the equilibrium curve $y_i = f(x_i)$. The vapor leaving the plate is always *richer in lightweight substances* than the liquid left behind. Thus, at the top of the column, light weights are found in the distillate, and at the bottom, heavier materials are recovered in the residue.

The absorption of a gaseous compound into a liquid may take place either adiabatically or else when the plates are cooled down, as happens during the synthesis of nitric acid.

Stripping consists of vaporizing a compound dissolved in a liquid (e.g. extraction of bromine from seawater). In order to do so, we bring an inert (non-soluble) gas into contact with the solution, the effect of which is to decrease the partial vapor pressure of the solute above the solution.

The methods discussed in this chapter enable us to determine the number of theoretical plates needed to separate out the components of a mixture to attain predefined levels of purity.

1.1.2. *Practical data*

In our discussion here, we shall use:

1) The saturating vapor pressures:

It is helpful to express these using Antoine's equation:

$$\pi(t) = A - \frac{B}{t + C} \quad (t \text{ in } °C)$$

2) The equilibrium coefficients:

By definition, the equilibrium coefficient of the component i is the ratio y_i/x_i of the molar fraction in the gaseous phase to the molar fraction in the liquid phase.

If we have a single equation of state for both phases, it will be sufficient to write that the fugacity of the component i has the same value in the two phases:

$$f_{Li} = \phi_{Li} x_i P_T = \phi_{Vi} y_i P_T = f_{Vi}$$

Thus, we have the following expression of the equilibrium coefficient E_i:

$$E_i = y_i / x_i = \phi_{Li} / \phi_{Vi}$$

ϕ_{Li} and ϕ_{Vi} are the fugacity coefficients of the component i in the liquid and in the vapor.

If we do not have an equation of state and if the gaseous phase is far from the critical conditions, we can express ϕ_{Vi} with the equation of the virial and deal with the liquid phase in a real solution by bringing into play the activity coefficient γ_i. We would then write:

$$f_{Vi} = \phi_{Vi} y_i P_T = \gamma_i x_i \pi_i = f_{Li}$$

P_T : total pressure of the system: Pa

π_i : saturating vapor pressure of the component i in the pure state: Pa

Therefore:

$$E_i = \frac{\gamma_i \pi_i}{\phi_{Vi} P_T}$$

3) Enthalpies:

The vapor enthalpy H^V and liquid enthalpy h^L of each component can be expressed by linear functions of the temperature (in the simplest cases) or by higher-degree polynomials. It must be remembered that the difference $(H^V - h^L)$ is the latent heat of vaporization which may be deduced from the saturating vapor pressure by Clapeyron's equation:

$$\Delta H = \frac{d\pi(t)}{dt} \times T \Delta V$$

ΔH: molar latent heat of vaporization: $J.kmol^{-1}$

ΔV: difference of the molar volumes of the vapor and the liquid: $m^3.kmol^{-1}$

t and T: temperatures in Celsius and Kelvin

The enthalpy of the gaseous phase will often be a weighted mean of the enthalpies of the components:

$$H^V = \sum_i H_i^V y_i$$

However, the enthalpy of the liquid phase must often include the excess enthalpy h^E.

The enthalpy of the liquid will then be:

$$h^L = \sum_i h_i^L x_i + h^E$$

1.1.3. Calculation methods presented in this chapter

Three methods are found here:

– the simple, graphical method advanced by McCabe and Thiele. This method is useful only for binary mixtures;

– the global method, which is used for the simulation of an existing column, regardless of the number of components. This method is also apt if lateral discharges are expected;

– the successive plate method, whereby the calculations are performed on the basis of the two extremities of the column. When convergence is reached, the results of the calculation of the feed plate are the same as when we start at one end or the other of the column. This method can be used to directly find the number of plates necessary.

Unlike the global method, the successive plate method is unable to take account of any lateral discharge.

1.2. McCabe and Thiele's method

1.2.1. *Hypotheses specific to McCabe and Thiele's method*

1) The sensible heats are discounted, and the excess enthalpies considered to be null.

2) The molar latent heats of state change (vaporization or liquefaction) are equal for the two components of the mixture.

It results from this that the vapor and liquid flowrates are *constant* along each of the two sections of the column, though on condition that we discount the sensible variations in heat.

1.2.2. *Equilibrium curves y = f(x)*

For certain binary mixtures, we may define a relative volatility α of the lightweight species A in relation to the heavy compound B:

$$\alpha = \left[\frac{y_A}{x_A}\right]\left[\frac{x_B}{y_B}\right] = \frac{y_A}{1-y_A} \times \frac{1-x_A}{x_A} \text{ so } y_A = \frac{\alpha x_A}{1+(\alpha-1)x_A} \text{ (where } \alpha > 1)$$

The equilibrium curve passes through the origin $(y_A = x_A = 0)$ and through the point $[y_A = x_A = 1]$. It is situated above the diagonal $y = x$ and deviates from it all the more when α is greater. If $\alpha = 1$, it is identical to that diagonal. If we accept the laws of ideality we can write:

$$Py_A = x_A \pi_A \text{ and } Py_B = x_B \pi_B \text{ and therefore } \alpha = \pi_A / \pi_B$$

π_A and π_B are the vapor pressures of the light species A and the heavy species B.

In most situations encountered in industry, the idea of relative volatility independent of the compositions is not appropriate. Thus, we need to use the equilibrium curve $y = f(x)$ or $x = g(y)$, determined experimentally.

Note that the experimental curves all pass through the origin and through the point $x = y = 1$. If an azeotrope exists, the equilibrium curve crosses the diagonal $y = x$ at the point corresponding to the composition of the azeotrope.

1.2.3. Material balances

The feed F splits the column into two sections. The accepted conventions dictate that we refer to the light species and, therefore, that the upper section be called the enriching section and the lower section be called the stripping (exhausting) section.

From the accepted hypotheses, it stems that, in each section, the downward liquid molar flowrate L is constant, and so too is the upward vapor molar flowrate V.

1) Enriching section (operating line):

Let us isolate a domain surrounding the top of the column and several plates. Write that the input is equal to the output. D is the flowrate of the material decanted into the condenser (the distillate):

$$V = L + D$$

More specifically, let the plates be numbered from the top to the bottom of a section. Plate number 1 is constituted by the condenser. The boundary of the domain runs between plates n and n + 1. Let us write that what enters the domain thus defined and what exits it are exactly the same:

$$Vy_{n+1} = Lx_n + Dx_D$$

If we eliminate V between these two equations, we find:

$$y_{n+1} = \frac{L}{L+D}x_n + \frac{D}{L+D}x_S$$

Let us introduce the reflux ratio R = L/D. We obtain the equation for a straight line in the plane x, y.

$$y_{n+1} = \frac{R}{R+1}x_n + \frac{x_D}{R+1}$$

This line is the operating line for enrichment. It passes through the two points:

$$[x = x_D, y = x_D] \text{ and } [x = 0, y = x_D/(R+1)]$$

2) Exhausting section:

The vapor and liquid flowrates are denoted V' and L'; let the plates be numbered from bottom to top. W is the flowrate of material deposited at the bottom of the column (the residue):

$$L' = V' + W$$

$$L'x_{m+1} = V'y_m + Wx_W$$

Thus, by eliminating L' and setting $\theta = V'/W$, we obtain the equation of the operating line for exhaustion:

$$x_{m+1} = \frac{\theta y_m}{1+\theta} + \frac{x_W}{1+\theta}$$

θ is the revaporization rate.

c) Thermal state of the feed:

At the point I where the operating lines meet, we must have:

$$Vy_I = Lx_I + Dx_D \qquad [1.1]$$

$$V'y_I = L'x_I - Wx_W \qquad [1.2]$$

Let us subtract these two equations from one another, term by term:

$$(V - V')y_I = (L - L')x_I + Dx_D + Wx_W \qquad [1.3]$$

Let τ be the molar fraction of the feed vaporized. We have:

$$V - V' = \tau F \quad \text{and} \quad L' - L = (1 - \tau)F$$

The overall balance of the lightweight species is written:

$$Dx_D + Wx_W = Fz$$

z is the feed's content in light species.

Equation [1.3] is then written:

$$y_I = \frac{\tau - 1}{\tau}x_I + \frac{z}{\tau} \qquad [1.4]$$

This equation defines the line of thermal state of the feed.

We can verify that the meeting point I between lines [1.1] and [1.2] satisfies equation [1.4], so the line of thermal state passes through the two points F and I (see Figure 1.1):

$$F[x = z, y = z]$$

$$I[x = x_I, y = y_I]$$

It is rare for the line of thermal state to intersect the equilibrium curve precisely at a point representative of a theoretical plate.

If the mixture fed in is two-phased, its enthalpy is expressed by:

$$H = \tau H^V + (1-\tau)h^L \text{ where } 0 < \tau < 1$$

For a superheated vapor, we have:

$$H = H^V v \text{ where } v > 1$$

For a supercooled liquid:

$$H = h^L \lambda \text{ where } \lambda < 1$$

Let us now find the corresponding value of τ:

$$H^V v = H^V \tau + h^L(1-\tau)$$

Hence:

$$\tau = \frac{H^V v - h^L}{H^V - h^L} > 1 \text{ for the superheated vapor}$$

Similarly, for the supercooled liquid:

$$h^L \lambda = H^V \tau + h^L(1-\tau)$$

Thus:

$$\tau = \frac{h^L(\lambda - 1)}{H^V - h^L} < 0 \text{ for the supercooled liquid}$$

This generalizes the use of the line of thermal state of the feed, but parameter τ has no physical significance outside of the interval $[0,1]$.

If $\tau = 0$, the result is that $x_I = z$ and the line of thermal state (equation 1.4) is vertical. If $\tau = 1$, then $y_I = z$ and the line is horizontal. If $\tau > 1$, the slope of the line is positive, and if $\tau < 0$, then the slope is positive as well. On the other hand, for a two-phase feed, the slope of the line of thermal state is negative (i.e. for $0 < \tau < 1$).

If the line of thermal state (which passes through I) intersects the equilibrium curve at a point E (see Figure 1.1), identical to the point representative of a plate, then the arrival of the feed on that plate will not result in either the vaporization or the liquefaction of the feed. Generally, this is not the case.

1.2.4. *Plotting of the tiers*

1) Enriching:

We suppose that the vapor coming from plate 1 (the highest plate in the column) is entirely condensed. For this vapor:

$$y_1 = x_D \quad V = (R+1)D$$

Similarly, for the liquid reflux reaching plate 1:

$$x_0 = x_D \qquad L = RD$$

The equation of the operating line is satisfied for plates 0 and 1:

$$y_1 = \frac{R}{R+1} x_0 + \frac{x_D}{R+1} = x_D$$

The composition of the liquid exiting plate 1 is given by the equilibrium curve:

$$x_1 = f(y_1)$$

The composition of the vapor y_2 is then given by:

$$y_2 = \frac{Rx_1}{R+1} + \frac{x_D}{R+1}$$

We obtain:

$$x_2 = f(y_2) \text{ etc.}$$

which is represented graphically in Figure 1.1, as follows:

From the point $y = x_D$ on the bisector, we draw a horizontal toward the equilibrium curve, so x_1. The operating line gives y_2 for the vertical with abscissa value x_1, and so on.

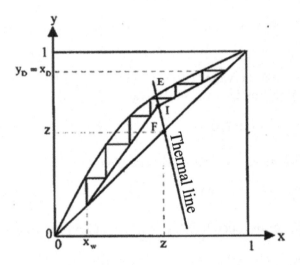

Figure 1.1. *Graphical construction of McCabe and Thiele*

2) Exhaustion:

We take the reboiling (revaporization) rate θ.

The residue decanted at the bottom of the column has the composition x_W. The rising vapor V' has the composition $y_0 = f(x_W)$.

The liquid arriving at the bottom of the column has the composition given by the equation of the operating line:

$$x_1 = \frac{\theta y_0}{1+\theta} + \frac{x_W}{1+\theta}$$

The vapor leaving plate no. 1 has the composition:

$$y_1 = f(x_1)$$

The composition of the liquid falling from plate 2 is:

$$x_2 = \frac{\theta y_1}{\theta+1} + \frac{x_W}{1+\theta}$$

Therefore:

$$y_2 = f(x_2) \text{ etc.}$$

which is represented graphically in Figure 1.1, as follows:

From the point $x = x_W$, we draw a vertical line which gives the composition $y_0 = f(x_W)$ of the vapor leaving the bottom of the column. The horizontal with the ordinate y_0 cuts the operating line at the point with abscissa x_1, which gives y_1 by the equilibrium curve $y_1 = f(x_1)$, etc.

1.2.5. *Overall material balance of the column*

The material balance is written:

$$F = D + W \text{ (kilomoles per second)}$$

The material balance relative to the volatile species is:

$$Fz = Dx_D + Wx_W$$

If, for example, we take a specific value for z, x_D and x_W, those two equations give us $w = W/F$ and $d = D/F$:

$$w = \frac{W}{F} = \frac{z - x_D}{x_W - x_D} \text{ and } d = \frac{D}{F} = \frac{z - x_W}{x_D - x_W}$$

Strictly speaking, it would be preferable to employ McCabe and Thiele's method, setting:

$$F = 1 \quad W = w \text{ and } D = d$$

1.2.6. *Overall heat balance for the column*

Consider:

Q_B : thermal power of the boiler: Watt

Q_C : thermal power of the condenser: Watt

C_W : molar specific heat capacity of the residue: $J.kmol^{-1}.°C^{-1}$

t_W : temperature at the bottom of the column (and therefore of the residue): °C

t_D : temperature of the distillate: °C

C_D : molar specific heat capacity of the distillate: $J.kmol^{-1}.°C^{-1}$

H_F : mean molar enthalpy of the feed: $J.kmol^{-1}$

$$H_F = \tau H_V + (1-\tau)h_L$$

where:

H_V : enthalpy of the vaporized fraction of the feed: $J.kmol^{-1}$

h_L : enthalpy of the liquid fraction of the feed: $J.kmol^{-1}$

τ : ratio of vaporization of the feed

The overall balance is then written:

$$FH_F + Q_B = DC_D t_D + WC_W t_W + Q_c$$

In practical terms, we set the flowrate of distillate D and the reflux ratio R (generally between 2 and 5).

In view of the operating pressure of the column and supposing we know the vapor pressures of the light and heavy species, we deduce the temperatures t_W and t_D.

The power of the condenser can be deduced from this:

$$Q_C = D(R+1)\Delta H_D$$

ΔH_D: latent heat of condensation of the lightweight species: $J.kmol^{-1}$

The overall heat balance gives us the power of the boiler.

$$Q_B = DC_D t_D + WC_W t_W + Q_C - FH_F$$

In practice, it is wise to increase the power Q_B by 10% in order to allow for the inevitable thermal losses.

1.2.7. Revaporization ratio of the boiler

It is tempting, if we know Q_B, to write:

$$V' = Q_B / \Delta H_w$$

ΔH_W : latent heat of vaporization of the heavy species

Hence:

$$\theta = V'/W$$

In reality, if we wish to maintain consistency with the hypotheses underpinning McCabe and Thiele's method, we need to operate differently.

Consider a domain encapsulating the bottom of the column and a few plates from the exhausting section. Write that the input is equal to the output:

$$V' + W = L' = L + (1-\tau)F = RD + (1-\tau)F$$

Thus, we have the reboiling rate:

$$\theta = \frac{V'}{W} = \frac{(R+1)D - \tau F}{W}$$

If we know τ, R and θ it is possible to plot the three lines representing the column's operation:

– the two operating lines;

– the line of thermal state of the feed.

We can, for instance, set x_D, R and τ, which determines the operating line of enriching and the line of thermal state of the feed. Thus, we have the point I and, if we set x_W, the operating line of exhaustion is determined (see Figure 1.1).

1.2.8. Regulation of continuous distillation

In order to regulate the vapor of heating, we base our reasoning on a temperature reading at the sensitive point of the column. The sensitive point corresponds to the point of inflection of the curve illustrating the variation in temperature as a function of the level in the column. This is the point at which the temperature changes most quickly as a function of vapor flowrate in the reboiler – i.e. its thermal power.

Incidentally, note that reboilers generally have pipes whose internal diameter is 2 cm. These pipes are vertical and 2 meters long. Their heat transfer coefficient is close to $500 \ W.m^{-2}.°C^{-1}$. The heating vapor is outside the pipes.

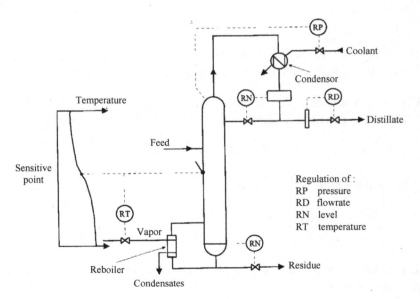

Figure 1.2. *Regulation in a distillation column*

Condensers have pipes which may be over two meters long. The internal diameter of these pipes is generally 2 cm. The cooling fluid circulates in the pipes. The presence of un-condensable gases greatly decreases the transfer coefficient, which may drop to 10 $W.m^{-2}.°C^{-1}$.

1.2.9. Number of plates for high purity of the light species

The equilibrium curve passes through the point $x = y = 1$ and has the slope m:

$$y - 1 = m(x - 1)$$

Thus:

$$y = mx + (1 - m)$$

The operating line has the slope $\dfrac{L}{V}$ and passes through the point $y = x = x_D$ (D as distillate).

$$y_n = \frac{L}{V} x_{n+1} + (1 - \frac{L}{V}) x_D$$

These two lines intersect at a point P outside of the square x between 0 and 1 and y between 0 and 1.

$$x_P = \frac{(1 - m) - (1 - \dfrac{L}{V}) x_D}{\dfrac{L}{V} - m}$$

The y and x values have an identical index, which is that of the point on the equilibrium curve which they characterize. This point is the image of a plate.

For reasons of proportionality, we see that:

1) on the operating line:

$$\frac{L}{V} = \frac{y_P - y_0}{x_P - x_1} = \frac{y_P - y_1}{x_P - x_2} = ...- = \frac{y_P - y_{n-1}}{x_P - x_n}$$

2) on the equilibrium line:

$$m = \frac{y_P - y_0}{x_P - x_0} = \frac{y_P - y_1}{x_P - x_1} = ... - = \frac{y_P - y_{n-1}}{x_P - x_{n-1}}$$

3) and, by dividing the fractions L/V by the fractions m and multiplying the fractions obtained by one another, we find:

$$\left[\frac{L}{Vm} \right]^N = \frac{x_P - x_0}{x_P - x_n}$$

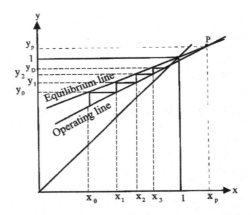

Figure 1.3. *McCabe and Thiele's plot (enriching end)*

The number of theoretical plates necessary, therefore, is:

$$N = \frac{Ln \left[\dfrac{x_P - x_0}{x_P - x_n} \right]}{Ln \left[\dfrac{L}{Vm} \right]}$$

EXAMPLE 1.1.–

How many theoretical plates would be needed to increase the purity of methanol from 0.9 to 0.9999? The impurity is water.

$$\frac{L}{V} = 12 / 20 = 0.6 \quad m = 0.45 \quad x_D = 0.9999 \quad x_0 = 0.9$$

$$x_P = \frac{0.55 - 0.4 \times 0.9999}{0.6 - 0.45} = 1.00027$$

$$N = \frac{Ln \dfrac{1.00027 - 0.9}{1.00027 - 0.9999}}{Ln \dfrac{0.6}{0.45}} = 19.5$$

Thus, we have 20 theoretical plates.

1.2.10. *Number of plates for a high degree of purity of the heavy species*

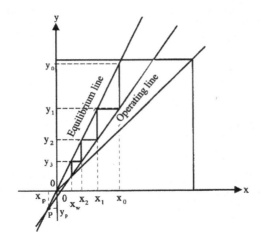

Figure 1.4. *McCabe and Thiele's plot (exhausting end)*

The equilibrium curve passes through the origin and has the slope E:

$$y = Ex$$

The operating line has the slope L'/V' and passes through the point $y = x = x_W$:

$$y_{n'} = \frac{L'}{V'} x_{n'+1} + (1 - \frac{L'}{V'}) x_W$$

Numbering proceeds from top to bottom. These two lines intersect at a point P outside of the square x between 0 and 1 and y between 0 and 1.

$$x_P = \frac{\left(\dfrac{L'}{V'} - 1\right)x_W}{\dfrac{L'}{V'} - E} \quad \text{and} \quad y_P = Ex_P$$

Similarly to what has been demonstrated for the enriching section:

$$\frac{L'}{V'} = \frac{y_1 - y_P}{x_0 - x_P} = \frac{y_2 - y_P}{x_0 - x_P} = \cdots\cdots = \frac{y_n - y_P}{x_{n-1} - x_P}$$

$$\frac{1}{E} = \frac{x_0 - x_P}{y_0 - y_P} = \frac{x_1 - x_P}{y_1 - y_P} = \cdots\cdots = \frac{x_{n-1} - x_P}{y_{n-1} - y_P}$$

$$\left[\frac{L'}{V'E}\right]^N = \frac{y_n - y_P}{y_0 - y_P}$$

The number of theoretical plates necessary, therefore, is:

$$N = \frac{Ln\left[\dfrac{y_n - y_P}{y_0 - y_P}\right]}{Ln\left[\dfrac{L'}{V'E}\right]}$$

EXAMPLE 1.2.–

How many theoretical plates would be needed to decrease the water content of a methanol solution from 0.18 to 0.0001?

$$\frac{L'}{V'} = 1.0444 \quad E = 3.72 \quad x_W = 10^{-4} \quad y_0 = 0.18$$

$$x_P = \frac{0.0444.10^{-4}}{1.0444 - 3.72} = -1.659.10^{-6}$$

$$y_P = -1.659.10^{-6} \times 3.72 = -6.1715.10^{-6}$$

$$N = \frac{Ln\left[\dfrac{10^{-4} + 1.659 \cdot 10^{-6}}{0.18 + 1.659 \cdot 10^{-6}}\right]}{Ln\left(\dfrac{1.0444}{3.72}\right)} = 5.88$$

Thus, we have 6 theoretical plates.

1.3. Global method (more than two components)

1.3.1. *Equations and unknowns*

Consider a column with N plates with indices j (for the condenser $j = 1$ and for the reboiler $j = N$) dealing with a mixture of c components with the indices i. We shall present the results found by Taylor and Edmister [TAY 69], which are explained by those authors themselves. Let us specify a number of additional matters.

We take the following variables:

– the feeds in each plate in terms of composition, temperature and flowrate, the fractions remaining after lateral discharge in either liquid or vapor form. These are the b_j and B_j, which we shall discuss later on. Obviously, these remaining fractions are between 0 and 1;

– the heats Q_j applied to each plate.

The goal is to determine the following values:

– the partial liquid and vapor flowrates of each component i exiting each plate j, so we have 2cN unknowns;

– the overall liquid and vapor flowrates exiting each plate, so we have 2N unknowns;

– the temperatures T_j of each plate, so we have N unknowns.

Thus, in total, we are looking for 2cN + 3N unknowns. For this purpose, we have the following relations:

– cN partial material balances;

– cN equilibrium relations;

– N overall material balances;

– N enthalpy balances;

– N boiling-point relations.

In total, then, we have $2cN + 3N$ relations.

We shall now examine each of these relations in turn.

1.3.2. Partial material balance

The partial material balance is the balance pertaining to a given component taken in isolation.

Obviously, the sum of the partial flowrates over all the components is equal to the total flowrate. Thus, the vapor leaving plate j is:

$$\sum_{i=1}^{c} v_{ji} = V_j \quad \text{with} \quad y_{ji} = \frac{v_{ji}}{V_j}$$

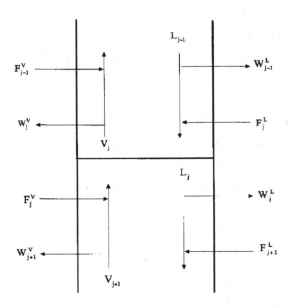

Figure 1.5. *Partial material balance of the component i*

Similarly, for the liquid leaving plate j:

$$\sum_{i=1}^{c} l_{ji} = L_j \ \text{ with } \ x_{ji} = \frac{l_{ji}}{L_j}$$

In addition, the following flowrates are discharged from plate j:

− vapor:

$$\sum_{i=1}^{c} w_{ji}^{V} = W_{j}^{V} \ \text{ with } \ y_{ji} = \frac{v_{ji}}{V_j} = \frac{w_{ji}^{V}}{W^{V}}$$

− liquid:

$$\sum_{i=1}^{c} w_{ji}^{L} = W_{j}^{L} \ \text{ with } \ x_{ji} = \frac{l_{ji}}{L_j} = \frac{w_{ji}^{L}}{W^{L}}$$

Around plate j, the balance of the component i is written:

$$l_{j-1,i} - w_{j-1,i}^{L} + f_{ji}^{L} + v_{j+1,i} - w_{j+1}^{V} + f_{ji}^{V} = v_{ji} + l_{ji}$$

f_{ji}^{L} and f_{ji}^{V} are the feeds of liquid and vapor to plate j.

Let us now define the absorption factor A_{ji} pertaining to the component i on the plate j by the relation:

$$A_{ji} = \frac{L_j}{E_{ji} V_j}$$

The partial equilibrium equations are written:

$$E_{ji} = \frac{y_{ji}}{x_{ji}} = \frac{v_{ji}}{V_j} \times \frac{L_j}{l_{ji}}$$

Consequently:

$$l_{ji} = A_{ji} v_{ji}$$

The balance equation of the component i becomes:

$$-\left[1-\frac{w_{j-1}^{L}}{L_{j-1}}\right]A_{j-1,i}v_{j-1,i}+(1+A_{ji})v_{ji}-\left[1-\frac{w_{j+1}^{V}}{v_{j+1}}\right]v_{j+1,i}=f_{ji}^{L}+f_{ji}^{V}$$

Let us set:

$$B_{j}=1-W_{j}^{V}/V_{j}\ \text{and}\ b_{j}=1-W_{j}^{L}/L_{j}$$

and also:

$$\phi_{ji}=f_{ji}^{L}+f_{ji}^{V}\quad(\text{here, }\phi_{ji}\text{ is not a fugacity coefficient})$$

The balance equation is then written:

$$-b_{j-1}A_{j-1,i}v_{j-1,i}+(1+A_{ji})v_{ji}-B_{j+i}v_{j+1,i}=\phi_{ji}$$

The term $b_{j-1}A_{j-1,i}v_{j-1,i}$ plays no part on the upper plate in the column.

Similarly, the term $B_{j+1}v_{j+1,i}$ does not play a part for the lower plate in the column.

Thus, we obtain the following system of partial material balance equations:

$$(1+A_{1i})v_{1i}-B_{2}v_{2i}=\phi_{1i}$$

$$-b_{1}A_{1i}v_{1i}+(1+A_{2i})v_{21}-B_{3}v_{3i}=\phi_{2i}$$

$$-b_{j-1}A_{j-1,i}v_{j-1,i}+(1+A_{ji})v_{ji}-B_{j+1}v_{j+1,i}=\phi_{ji}$$

$$-b_{N-2}A_{N-2,i}v_{N-2,i}+(1+A_{N-1,i})v_{N-1,i}-B_{N}v_{Ni}=\phi_{N-1,i}$$

$$-b_{N-1}A_{N-1,i}v_{N-1,i}+(1+A_{Ni})v_{Ni}=\phi_{Ni}$$

Similarly, if we define the stripping factor by:

$$S_{ji}=E_{ji}\frac{V_{j}}{L_{j}}$$

we obtain a system of equations equivalent to the previous one:

$$(1+S_{1i})l_{1i} - B_2 S_{2i} l_{2i} = \phi_{1i}$$

$$-b_1 l_{1i} + (1+S_{2i})l_{2i} - B_3 S_{3i} l_{3i} = \phi_{2i}$$

$$-b_{j-1} l_{j-1,i} + (1+S_{ji})l_{ji} - B_{j+1} S_{j+1,i} l_{j+1,i} = \phi_{ji}$$

$$b_{N-2} l_{N-2,i} + (1+S_{N-1,i})l_{N-1,i} - B_N S_{Ni} l_{Ni} = \phi_{N-1,i}$$

$$-b_{N-1} l_{N-1,i} + (1+S_{Ni})l_{Ni} = \phi_{Ni}$$

1.3.3. *Solutions to partial balance equations (compound i)*

For the heavy components (small value of E_{ji}):

$$S_{ji} = E_{ji} \frac{V_j}{L_j} \quad \text{and} \quad \prod_{j=1}^{N} S_{ji} \le 1$$

$$\sigma_{0i} = 1$$

$$\sigma_{1i} = 1 + S_{1i}$$

$$\sigma_{ji} = \sigma_{j-1,i}(1+S_{ji}) - \sigma_{j-2,i} b_{j-1} B_j S_{ji} \qquad 2 \le j \le N$$

$$l_{Ni} = \frac{\phi_{Ni}\sigma_{N-1,i} + \sum_{q=1}^{N-1}\left[\left[\prod_{t=q}^{N-1} b_t\right]\phi_{qi}\sigma_{q-1,i}\right]}{\sigma_{Ni}}$$

$$l_{ji} = \frac{\phi_{ji}\sigma_{j-1,i} + \sum_{q=1}^{j-1}\left[\left[\prod_{t=q}^{j-1} b_t\right]\phi_{qi}\sigma_{q-1,i}\right] + l_{j+1,i} S_{j+1,i} B_{j+1}\sigma_{j-1,i}}{\sigma_{ji}} \qquad N-1 \ge j \ge 2$$

$$l_{1i} = \frac{\phi_{1i} + l_{2i} S_{2i} B_2}{\sigma_{1i}}$$

For the light components (large value of E_{ji}):

$$A_{ji} = \frac{L_j}{E_{ji}V_j} \text{ and } \prod_{j=1}^{N} A_{ji} \le 1$$

$$\alpha_{N+1,i} = 1$$

$$\alpha_{Ni} = 1 + A_{Ni}$$

$$\alpha_{ji} = \alpha_{j+1,i}(1 + A_{ji}) - B_{j+1}b_j A_{ji}\alpha_{j+2,i} \qquad\qquad N-1 \ge j \ge 1$$

$$v_{1i} = \frac{\phi_{1i}\alpha_{2i} + \sum_{q=2}^{N}\left[\left[\prod_{t=2}^{q} B_t\right]\phi_{qi}\alpha_{q+1,i}\right]}{\alpha_{1i}}$$

$$v_{ji} = \frac{\phi_{ji}\alpha_{j+1,i} + \sum_{q=j+1}^{N}\left[\left[\prod_{t=j+1}^{q} B_t\right]\phi_{qi}\alpha_{q+1,i}\right] + A_{j-1,i}b_{j-i}v_{j-i,i}\alpha_{j+1,i}}{\alpha_{ji}} \qquad 2 \le j \le N-1$$

$$v_{Ni} = \frac{\phi_{Ni} + \alpha_{N-1,i}b_{N-1}v_{N-1,i}}{\alpha_{Ni}}$$

1.3.4. *Overall material balances*

Let F_j^L and F_j^V be the feeds of liquid and vapor onto plate j. We know that W_j^V and W_j^L are the discharges of vapor and liquid on the plate. The balance for plate j is written:

$$L_{j-1} - W_{j-1}^L + F_j^L + V_{j+1} - W_{j+1}^V + F_j^V = L_j + V_j$$

The system of equations can be rendered explicit for the V_j values:

$$V_1 - V_2 \qquad\qquad = F_1^L + F_1^V - W_2^V$$

$$V_2 - V_3 \qquad\qquad = F_2^L + F_2^V + L_1 - L_2 - W_1^L - W_3^V$$

$$.$$
$$.$$
$$.$$

$$V_j - V_{j+1} \qquad\qquad = F_j^L + F_j^V + L_{j-1} - L_j - W_{j-1}^L - W_{j+1}^V$$

$$.$$
$$.$$
$$.$$

$$V_{N-1} - V_N \quad = F_{N-1}^L + F_{N-1}^V + L_{N-2} - L_{N-1} - W_{N-2}^L - W_N^V$$

$$V_N \quad = F_N^L + F_N^V + L_{N-1} - L_N - W_{N-1}^L$$

1.3.5. *Heat balances*

For plate j, the balance of inputs and outputs is written:

$$\sum_{i=1}^{c} l_{j-1,i} h_{j-1,i} - \sum_{i=1}^{c} w_{j-1,i}^L h_{j-1,i} + \sum_{i=1}^{c} f_{ji}^L h_{ji}^F +$$

$$\sum_{i=1}^{c} v_{j+1,i} H_{j+1,i} - \sum_{i=1}^{c} w_{j+1,i}^V H_{j+1,i} + \sum_{i=1}^{c} f_{ji}^V H_{ji}^F + Q_j$$

$$= \sum_{i=1}^{c} l_{ji} h_{ji} + \sum_{i=1}^{c} v_{ji} H_{ji} \qquad\qquad 1 \le j \le N$$

The quantity Q_j is the thermal power applied to plate j and counted positively if that plate receives heat. In a distillation column, Q_N will be positive (reboiler) and Q_1 negative (condenser).

The above balance can be written:

$$b_{j-1} \sum_{i=1}^{c} l_{j-1,i} h_{j-1,i} + B_{j+1} \sum_{i=1}^{c} S_{j+1,i} H_{j+1,i} l_{j+1,i} + \sum_{i=1}^{c} f_{ji}^L h_{ji}^F + \sum_{i=1}^{c} f_{ji}^V H_{ji}^F + Q_j$$

$$= \sum_{i=1}^{c} l_{ji} h_{ji} + \sum_{i=1}^{c} S_{ji} l_{ji} H_{ji}$$

In these equations, the h values represent the enthalpies of the liquids and the H those of the vapors.

The thermal power of the reboiler is deduced from that of the condenser by finding an overall balance for the column.

$$Q_N = Q_1 - \sum_{j=1}^{n} \sum_{i=1}^{c} (f_{ji}^L h_{ji}^F + f_{ji}^V H_{ji}^F - w_{ji}^L h_{ji} - w_{ji}^V H_{ji})$$

An estimation of the right-hand side of this equation enables us to evaluate Q_N.

The power of the condenser results from the choice of the reflux ratio R of the column.

$$Q_1 = (R+1) \sum_{i=1}^{c} w_{1,i}^L (H_{2,i} - h_{1,i})$$

$$Q_1 = (R+1) D (H_2 - h_1)$$

D is the flowrate of the distillate.

Stripping operations generally take place without heat exchange with the outside world. On the other hand, this is not always the case with absorptions, because it may be that the plates are cooled.

1.3.6. *Boiling-point relations*

On each plate, it is sufficient, by a classic calculation of liquid–vapor equilibrium, to solve the equation:

$$\sum_{i=1}^{c} E_{ji} x_{ji} = 1$$

We could use the tangent method. Remember that the E_{ji} depend on the temperature T_j.

With these relations, we are able to determine the temperatures of the plates.

1.3.7. *Global solution method*

We take linear initial profiles along the column for the temperatures T_j and liquid flowrates L_j. We then calculate any lateral discharges W_j:

$$W_j^L = L_j - b_j L_j$$

1) The overall balances give the V_j. From this, we deduce the W_j^V by:

$$W_j^V = V_j - B_j V_j$$

We repeat this procedure with the overall balances until the W_j^V no longer vary.

2) With the hypothesis of ideal behavior accepted, we calculate the A_{ji} and S_{ji}.

3) The partial material balances give us the v_{ji} and l_{ji}. From this, we deduce the compositions:

$$x_{ji} = \frac{l_{ji}}{\sum_{i=1}^{c} l_{ji}} \text{ and } y_{ji} = \frac{v_{ji}}{\sum_{i=1}^{c} v_{ji}}$$

4) The T_j are then calculated by N boiling-point equations (see section 1.3.6).

5) We then evaluate the enthalpies of the liquid- and vapor phases.

6) By combining the N heat balances and the N overall material balances, we calculate the $2\,N$ unknowns L_j and V_j by the Gauss–Jordan elimination method.

We go back to step 2 but, this time, we have composition profiles which enable us to evaluate the equilibrium coefficients E_{ji} in the hypothetical case of non-ideality.

The calculation is halted when the relative precision in terms of the liquid and vapor flowrates is 1‰ and 0.001°C on the temperatures T_j.

1.4. Successive plates method

1.4.1. General

To define a distillation column without lateral discharge, we define certain data of composition in the distillate and in the residue. The feed is known in terms of flowrate, composition and temperature. Here, we shall discuss certain points of the procedure put forward by [WUI 65].

The calculations are performed starting from the two ends of the column and working towards the feed.

1.4.2. Flowrate and composition of the distillate and of the residue

All the components must be specified either in the distillate D or in the residue W. Let us number the components whose specifications s_{wi} are given (in molar fractions) in the residue from 1 to r, and from $r+1$ to c the components whose specifications s_{di} are given in the distillate and set:

$$S_w = \sum_{i=1}^{r} s_{wi} \text{ and } S_d = \sum_{i=r+1}^{c} s_{di}$$

Suppose that the feed $F = W + D$ is split according to the specifications. For the component i specified in the residue:

$$Fz_i = Ws_{wi} + (F - W)z_{di} \text{ with } \sum_{i=1}^{r} z_{di} + S_d = 1$$

Thus:

$$F_w = F\sum_{i=1}^{r} z_i = WS_w + D\sum_{i=1}^{r} z_{di} = WS_w + (F - W)(1 - S_d)$$

Therefore:

$$W = \frac{F(1-S_d)-F_w}{1-S_d-S_w} \text{ and, similarly: } D = \frac{F(1-S_w)-F_d}{1-S_d-S_w}$$

The solution is indeterminate if $(1-S_d-S_w)=0$. We shall not examine the case in which all the components are specified in the same effluent – W or D – because to do so we would need to isolate each component and then make the desired mixture. However, it may happen that $S_w+S_d \geq 1$, but without S_w or S_d being equal to 1 or 0. To avoid this situation, we simply need to specify the content levels of *impurities* in D and W because, by nature, the impurity levels are much less than 1.

As regards the non-specified molar fractions:

$$z_{di} = \frac{Fz_i - Ws_{wi}}{F-W} \text{ and } z_{wi} = \frac{Fz_i - Ds_{di}}{F-D}$$

1.4.3. *Overall heat balance*

This balance is written by equaling the incomings and outgoings.

$Q_R + Q_F + Q_C = Q_D + Q_W$ (here, all the Q values are positive and expressed in watts)

The meaning of the indices is:

R: reboiler   F: feed   D: distillate

C: condenser   W: residue

The heat balance is useful in calculating the thermal power of the reboiler when we have determined that of the condenser.

Remember that the heat carried by a fluid mixture A (whose fraction L_A is liquid), i.e. its enthalpy, is:

$$Q_A = AH^E + A\sum_{i=1}^{c}\left[L_A x_{Ai} h_{Ai} + (1-L_A)y_{Ai}H_{Ai}\right]$$

In general, we overlook the excess enthalpy H^E.

The thermal power of the condenser is

$$Q_C = (R+1)D(H_D - h_D)$$

H_D is the enthalpy of the distillate in the gaseous state at its dew point and, for total condensation, h_D is that of the liquid distillate at its boiling point.

1.4.4. *Calculation of the plate temperatures*

Consider the upper section of the column. The plates are numbered from top to bottom, with 1 being the plate situated immediately beneath the condenser. Consider a domain encapsulating the condenser and plates 1 to j. For that domain, the balance of component i is written:

$$V_{j+1}y_{j+1,i} = L_j x_{ji} + Dx_{di} \qquad [1.5]$$

Let us multiply this equation by $H_{j+1,i}$ – i.e. the partial enthalpy of the component i in the vapor phase V_{j+1} – and sum in terms of i:

$$V_{j+1}\sum_{i=1}^{c} y_{j+1,i} H_{j+1,i} = L_j \sum_{i=1}^{c} x_{ji} H_{j+1,i} + D\sum_{i=1}^{c} z_{di} H_{j+1,i}$$

In addition, the heat balance is written:

$$V_{j+1}\sum_{i=1}^{c} y_{j+1,i} H_{j+1,i} = L_j \sum_{i=1}^{c} x_{ji} h_{ji} + Q_D + Q_C$$

By equaling the right-hand sides of these two equations, we find:

$$\varphi_{1j} = \frac{L_j}{D} = \frac{(Q_D + Q_C)/D - \sum_{i=1}^{c} z_{di} H_{j+1,i}}{\sum_{i=1}^{c} x_{j,i}(H_{j+1,i} - h_{j,i})}$$

However, we know that:

$$x_{j+1,i} = \frac{y_{j+1,i}}{E_{j+1,i}} \quad \text{and that:} \quad \sum_{i=1}^{c} x_{j,i} = 1$$

By dividing equation [1.5] by $E_{j+1,i}$ and summing in terms of i, we find:

$$\varphi_{2j} = \frac{L_j}{D} = \frac{1 - \sum_{i=1}^{c} \dfrac{z_{di}}{E_{j+1,i}}}{\sum_{i=1}^{c} \dfrac{x_{j,i}}{E_{j+1,i}} - 1}$$

Here, let us introduce the function: $R_j = \varphi_{1j} - \varphi_{2j}$.

The temperature T_{j+1} must render the function R_j equal to zero by way of the enthalpies $H_{j+1,i}$ and the equilibrium ratios $E_{j+1,i}$.

The denominator of φ_{2j} becomes zero for a temperature $T_{r,j}$, which is the dew point of a fictitious vapor with the composition x_{ji}. This temperature is greater than T_{j+1}. Indeed, the composition $x_{j,i}$ is less rich in light species than that composition $y_{j+1,i}$ because the light species are discharged with the distillate. In addition, the plate j + 1 is situated beneath plate j, and hence T_{j+1} is greater than T_j. Thus:

$$T_j < T_{j+1} < T_{r,j} \quad \text{(therefore, } T_j \text{ and } T_{r,j} \text{ are two limits for } T_{j+1})$$

Furthermore, φ_{1j} decreases with T_{j+1} and φ_{2j} grows with T_{j+1}. In other words, R_j is a *monotonic* decreasing function of T_{j+1}.

More generally, let $T_{j+1}^{(n-1)}$ and $T_{j+1}^{(n-2)}$ be two limits encapsulating T_{j+1}. We calculate $R_j = \overline{R}_j$ for $\overline{T}_{j+1}$, which is the arithmetic mean of those two limits. We eliminate the limit $T_{j+1}^{(x)}$ for which R_j has the same sign as $\overline{R}_j$, and replace it with $\overline{T}_{j+1}$. Thus, we obtain a narrower interval for T_{j+1}. The calculation is halted when $\overline{T}_{j+1}$ varies by less than 0.001°C between two operations.

For the lower section of the column (exhaustion of volatile species, which is tantamount to enriching in heavy species), the plates are numbered from bottom to top, with 1 being the plate situated just above the bottom of the column. The equations are:

$$L_{k+1}x_{k+1,i} = L_k y_{ki} + Wz_{wi}$$

$$L_{k+1}\sum_{i=1}^{c}x_{k+1,i}h_{k+1,i} = V_k \sum_{i=1}^{c}y_{ki}H_{ki} + Q_W - Q_R$$

From this, we derive:

$$\varphi_{1k} = \frac{V_k}{W} = \frac{\displaystyle\sum_{i=1}^{c}z_{wi}h_{k+1,i} - (Q_W - Q_R)/W}{\displaystyle\sum_{i=1}^{c}y_{ki}(H_{ki} - h_{k+1,i})}$$

$$\varphi_{2k} = \frac{V_k}{W} = \frac{1 - \displaystyle\sum_{i=1}^{c}z_{wi}E_{k+1,i}}{\displaystyle\sum_{i=1}^{c}y_{ki}E_{k+1,i} - 1}$$

$$R_k = \varphi_{1k} - \varphi_{2k}$$

R_k is a *monotonic* increasing function of T_{k+1}.

$$T_{b,k+1} < T_{k+1} < T_k$$

As we did for the upper section, we shall use the mean method (which is also known as the "dichotomy method").

1.4.5. *Compositions on the plates*

With regard to the upper section, knowing T_{j+1} gives us the value of L_j/D. The overall and partial material balances yield V_{j+1} and the $y_{j+1,i}$. Remember that these balances pertain to a domain encapsulating the condenser and plates 1 to j. The equilibrium relations $(x = y/E)$ give the $x_{j+1,i}$. For the lower section, T_{k+1} gives us V_k, so L_{k+1} and $x_{k+1,i}$ and finally $y_{k+1,i}$.

Generally, the ratios at equilibrium E depend on the compositions. Therefore we need to operate step-wise, taking, say, the initial value of E as:

$$E_i^{(0)} = \pi_i / P$$

π_i is the vapor pressure of the component i at T_{j+1} (or T_{k+1}) and P is the pressure.

1.4.6. *Consistency between the two sections*

It is possible to calculate the feed plate by starting either at the top or at the bottom. If the compositions of the distillate and the residue are correct, the compositions found for that plate must be the same by one method or the other.

If this is not the case, it is necessary to correct the specifications of the components at the top and at the bottom.

Let x_{fdi} and x_{fwi} represent the compositions of the liquid of the feed plate calculated respectively from top down and from bottom up.

1) $r + 1 \leq i \leq c$

The specification of the component has been defined at the top. The new value of the specification will be:

$$s_{di}^{(n+1)} = s_{di}^{(n)} \sqrt{\frac{x_{fwi}}{x_{fdi}}}$$

2) $1 \leq i \leq r$

The specification of the component has been defined at the bottom. The new value of the specification will be:

$$s_{wi}^{(n+1)} = s_{wi}^{(n)} \sqrt{\frac{x_{fdi}}{x_{fwi}}}$$

We have introduced the square roots to decrease the amplitude of the correction and prevent oscillations.

Also, in Wuithier [WUI 65], readers will find another way to correct the compositions of the distillate and the residue. However, it must not be forgotten that consistency with the feed plate is impossible to obtain if the specifications involve the crossing of an azeotrope. In this situation, only the global method will serve, but neither will this method enable us to cross the azeotrope.

1.5. Conclusion

1) If we need to design a simple column separating two components, we use the successive plate method. Having determined the specifications, we start at the top and the bottom, *a priori* taking a fairly high number of plates (at least 15) for each section. The engineer responsible will then examine the temperatures and compositions on each section and choose, as the feed plate, that for which the two calculations yield the closest results. This immediately gives us the number of plates in each section. We then merely need to employ the procedure of consistency between the two sections.

2) We may also seek to design a column with lateral discharge to isolate a compound that is present only in a low quantity in the feed. We first use the successive plate method as explained above, and once consistency between the two sections has been obtained, we look to see whether, along the length of the column, there is a maximum of concentration (a concentration "center") for the product we wish to discharge. Having chosen the discharge plate, we apply the global method.

3) The global method can be used manually (i.e. using a calculator) if the number of plates and the number of components are limited and if, more importantly, the equilibria are ideal. In general, the global method requires a computer with scientific precision (128 significant binary figures for each number).

4) The equivalence of a theoretical plate is approximately 1.4 to 1.7 real plates or, which is the same thing, a real plate is equivalent to 0.6 to 0.7 theoretical plates. Traditionally, though, numerous practitioners agree that the *equivalence of a real plate is 0.5 theoretical plates*.

1.6. Choice of type of column

For absorption or stripping distillation columns, the most commonly-used plates are perforated plates, because they are cheapest to make and of well-established design.

However, when the gaseous flowrate is very low, the liquid would pass through the holes instead of through the outlet(s). We then need to turn to bubble-cap plates, made with bubble caps of 10 cm nominal diameter.

When we want a slight drop in pressure on the side of the gas, we need to use a packed bed. Indeed, in this type of column, the gas only brushes against the liquid and does not pass through it. An additional advantage to these columns is how well they are suited to the treatment of corrosive fluids. Indeed, it is not overly costly to make the vessel out of a noble alloy and use ceramic or graphite as the packing.

Design and Performances of Gas–Liquid Perforated Plates

2.1. Geometry of the plate

2.1.1. *Advantage to using perforated plates*

Perforated plates are cheaper to make than bubble-cap plates, for which the calculations are given by Bolles [BOL 56] and which, today, are used only for very low gaseous flowrates because they do not present any danger of weeping.

2.1.2. *Diameter of the column*

Treybal [TRE 80] indicates that the value of the parameter $V_G\sqrt{\rho_G}$ must lie between 0.7 and 2.2 for proper operation.

V_G: in empty columns velocity of the vapor: $m.s^{-1}$

For an estimation of the column's diameter, we shall write:

$$V_G\sqrt{\rho_G} = \frac{1}{2}(0.7 + 2.2) = 1.45$$

i.e.:

$$V_G = \frac{1.45}{\sqrt{\rho_G}}$$

Thus, the diameter of the column is:

$$D_C = \sqrt{\frac{4Q_V}{\pi V_G}}$$

EXAMPLE 2.1.–

$$\rho_G = 1.25 \ kg.m^{-3} \quad Q_V = 11.66 \ m^3.s^{-1}$$

$$V_G = \frac{1.45}{\sqrt{1.25}} = 1.30 \ m.s^{-1}$$

The diameter of the column is:

$$D_C = \sqrt{\frac{4 \times 11.66}{\pi \times 1.30}} = 3.39 \ m$$

The section of the column is:

$$A_C = \frac{\pi \times 3.39^2}{4} = 9 \ m^2$$

2.1.3. Design of the downcomers

The liquid–vapor mixture disappears by separation of the bubbles and the clear liquid. The bubbles come together by coalescence, giving rise to larger bubbles, whose rate of ascension is sufficient for them to burst when they reach the surface. In order to evaluate the foaming nature of a liquid, we merely need to fill one third of a test tube with it and shake it hard for five seconds in the axial direction.

In order to do this, coalescence requires a sufficient time of stay, just like in a chemical reactor. This residence time is longer when the foam is stable.

Type of liquid	Residence time τ of the liquid supposed to be degassed (seconds)
Light hydrocarbons (non-foamy)	2–3
Heavy hydrocarbons (moderately foamy)	4–5
Glycols and amines (highly foamy)	6–7

Table 2.1. *Speed of bubbles*

The volume of the downcomer can be deduced from this:

$$\Omega_D = Q_L \tau$$

However, a downcomer is more complicated than a chemical reactor. Indeed, the gas bubbles only disappear on the free surface of the foam. The downcomer then behaves like an inverted decanter in which the discontinuous phase gathers in the upper part. According to the decanter theory, the downcomer must provide the foam with a sufficient "decantation" surface A_D where:

$$A_D = \frac{Q_L}{V_B}$$

Q_L : flowrate of clear liquid: $m^3.s^{-1}$

V_B : rate of ascension of the bubbles ("decantation" rate): $m.s^{-1}$

The common values of the rate V_B are distributed as given in Table 2.2.

Type of liquid	Decantation rate $(m.s^{-1})$
Non-foamy	0.15
Moderately foamy	0.10
Highly foamy	0.07

Table 2.2. *Speed of bubbles*

In Appendix 3, readers will find a proposal of a standardized foaming test.

It is useful to know the *angle at the center 2 θ* blocked by the dam. For this purpose, we write that the section of the downcomer is the difference between the surface of the sector intercepted by the dam and the surface of the triangle delimited by the cord (see Figure 2.1):

$$A_D = \frac{R^2}{2}(2\theta - 2\sin\theta\cos\theta)$$

Thus: $A_D = \dfrac{D_C^2}{8} \varphi$ where: $\varphi = 2\theta - \sin 2\theta$

φ is an increasing function of θ. Remember that $1° = 0,017453\,\text{rad}$. The minimum value of the space between plates is:

$$S_{P\min} = \dfrac{\Omega_D}{A_D}$$

EXAMPLE 2.2.–

Moderately-foamy liquid

$Q_L = 0.02\ \text{m}^3.\text{s}^{-1}$ $D_C = 3.39\ \text{m}$

$V_B = 0.10\ \text{m.s}^{-1}$ $\tau = 4\ \text{s}$

$\Omega_D = 0.02 \times 4 = 0.08\ \text{m}^3$

$A_D = \dfrac{0.02}{0.10} = 0.2\ \text{m}^2$

$\varphi = \dfrac{8 \times 0.2}{3.39^2} = 0.139$

$\theta = 0.478\,\text{rad} = 27.39°$

$S_{P\min} = 0.08 / 0.2 = 0.4\ \text{m}$

Let us proceed by successive tests:

θ	0.45	0.47	0.48	0.478
φ	0.117	0.133	0.141	0.139

2.1.4. Possible configurations for the downcomer

The most typical type of spillway is the segmented downcomer, as illustrated below.

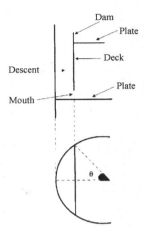

Figure 2.1. *Conventional downcomer (one pass)*

In general, the lower outlet from the deck is 10 mm lower than the height of the dam at the outlet from the plate. Additionally, the free height of the mouth for the passage of the liquid entering onto the plate will always be greater than 5 mm.

The dams at the entrance to the plate prevent unwanted weeping of the liquid, but they are inadvisable for viscid liquids. Their height must be equal to the height of the mouth. They cause a 20% increase of the pressure drop for the liquid entering onto the plate, on condition that the section available to the liquid is constant throughout its path.

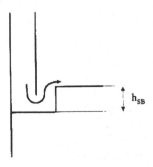

Figure 2.2. *Lowered guard*

Lowered liquid guards are always watertight and give the liquid and upward motion, which prevents weeping at the entrance to the plate.

The sink depth is 100 mm. The passage area available to the liquid must remain constant for all changes in direction.

At the exit from the plate, the downcomer contains a dam which is extended by an apron. When the diameter of a column increases, the length of the dam increases proportionally to the diameter, whilst the liquid flowrate increases with the square of the diameter.

Consequently, the line load on the dam quickly reaches the limiting value of $0.025 \ m^3.s^{-1}.m^{-1}$. Therefore, we need to increase the number of passes.

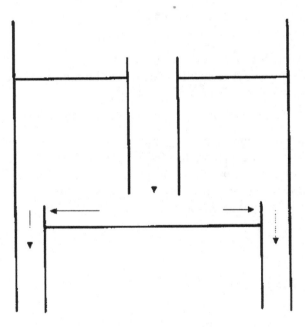

Figure 2.3. *Two-pass spillway*

Conversely, for small columns such as those encountered in pilots, the tube downcomer may be envisaged.

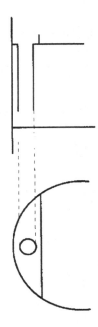

Figure 2.4. *Tube downcomer*

2.1.5. *Dam length and number of passes*

The dam length for a single pass is:

$$L_B = D_C \sin \theta$$

A single pass is justified if the liquid load at the dam is:

$$\frac{Q_L}{L_B} \leq 0.02 \, \mathrm{m^2.s^{-1}}$$

Otherwise, multiple passes are necessary, though it should be observed that the acceptable number of passes is limited. According to Economopoulos [ECO 78], we must have:

$$N_P \leq N_{P\,max} = 1.1 \; D_C \text{ (take the higher integer)}$$

If we wish to prevent the liquid taking preferential paths and some of this liquid is in weak contact with the gaseous phase, we must have:

$$\frac{L_B}{D_C} \geq 0.4$$

When the liquid load at the dam is less than $0.001 \ m^2.s^{-1}$, we must put in place a sawtoothed crenellated dam.

EXAMPLE 2.3.–

$$\theta = 0.478 \, rad \quad D_C = 3.39 \ m \quad Q_L = 0.02 \ m^3.s^{-1}$$

$$L_B = 3.39 \times \sin 0.478$$

$$L_B = 1.56 \ m$$

$$Q_L / L_B = 0.02/1.56 = 0.0128 < 0.02$$

We may content ourselves with a single pass, noting that the diameter of the column would allow for up to:

$$N_{P\,max} = 1.1 \times 3.39 = 3.7 \ \text{which represents four passes}$$

In addition, we can verify that:

$$\frac{L_B}{D_C} = \frac{1.56}{3.39} = 0.46 > 0.4$$

NOTE.– [WUI 72] gives additional information about downcomers downways.

2.1.6. *Active area*

It is necessary to leave an area on the plate without holes in, to take account of the rivets or welding that ensure the plate's rigidity and fixation. Similarly, we must allow for a calm zone, which therefore is not holed at the exit and entry to the spillways. We shall accept that these surfaces are equivalent to a circular band whose width is 3% of the diameter of the column. The area of the non-holed dead zone is therefore:

$$A_M = 0.03\pi D_C^2$$

The active area can be deduced from this:

$$A_A = A_C - A_M - \sum_{k=1}^{d} A_{Dk}$$

The sum of the A_{Dk} is the sum of the inlet and outlet areas of the downcomers on the plate. The term A_C is the area of the section of the column:

$$A_C = \frac{\pi D_C^2}{4}$$

EXAMPLE 2.4.–

The plate is single-pass.

$$A_C = 9 \text{ m}^2 \quad A_D = 0,2 \text{ m}^2 \quad D_C = 3.39 \text{ m}$$

$$A_A = 9 - 0.03\pi 3.39^2 - 2 \times 0.2$$

$$A_A = 7.52 \text{ m}^2$$

2.1.7. *Characteristics of holes*

The parameters defining a set of holes are:

– the diameter of the holes;

– the step between the holes;

– the thickness of the sheet metal.

1) In industrial practice, hole diameter varies from 5 to 15 mm.

Holes with a large diameter may be less costly than others. Indeed, they are more difficult to block and easier to clean than holes of a small diameter because, at constant perforated section, their total perimeter is smaller.

However, according to certain authors, when the velocity of the gas is high, or else when the diffusion on the side of the gas is difficult (high Schmidt number), it is preferable to use holes of moderate diameter.

The effectiveness of the plate depends on its operating regime. If we choose the foam regime, the vapor has little kinetic energy and, to encourage its dispersion in the liquid, many authors have suggested that numerous holes with small diameter (say, 3 mm, for instance) would lead to a large interfacial area.

On the other hand, in the jet regime, the dispersion of the liquid into droplets depends primarily on the vapor's kinetic energy (which is high) and the diameter of the holes is much less important. For a given perforated fraction ($\varphi = 0,1$, which is 10% of the active area, for example), it is cheaper to make holes of a larger diameter, of a restricted number. A diameter of 1 cm works well.

2) The second parameter to consider is the step p of the holes. Habitually, we define it by the ratio p/d_T of the step to the diameter of the holes.

In general:

$$2.5 < \frac{p}{d_T} < 4$$

When the ratio $x = p/d_T$ approaches its lower bound, the plate reaches its maximum "flexibility", meaning that its effectiveness is maintained at an acceptable level for low values of the gaseous flowrate. Indeed, in these conditions, the volume of the liquid separating two holes is small. Therefore, it is not necessary for the gas to vigorously stir the liquid in order for the exchange of material to be correct.

3) Knowing the ratio $x = p/d_T$, it is possible to determine the fraction ϕ of the active surface A_A which is covered by the holes – i.e. the degree of piercing. In general, this fraction is somewhere between 5 and 15%. Let A_T be the surface area of the holes.

For a triangular step:

$$\phi = \frac{A_T}{A_A} = \frac{\pi}{2\sqrt{3}x^2}$$

For a square step:

$$\phi = \frac{A_T}{A_A} = \frac{\pi}{4x^2}$$

The thickness e_p of the plate is the parameter which, along with the degree of piercing ϕ, characterizes the holes from the point of view of the pressure drop. In general, the thickness of the sheet metal chosen is 2 to 3 mm for stainless steel and noble alloys and 3 to 5 mm for soft steel.

For holes made in a thin sheet of metal (2 mm), the section of the gas jet is again reduced at the output from the hole, and the pressure recovers above the plate in a single step. On the other hand, if the sheet is thick (5 mm), an initial recovery of pressure takes place within the thickness of the metal, and a second recovery above the plate. However, the pressure recovery is better if it takes place in two steps.

2.1.8. *Plate of large diameter*

When the diameter of the plate is greater than 2 m, then on both sides of the liquid flow, zones may appear where the liquid rotates and does not renew as shown in Figure 2.5.

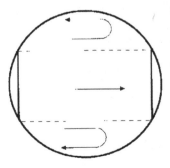

Figure 2.5. *Dead zones*

The existence of these true dead zones can seriously affect the effectiveness of the plate, particularly if the dams are short in length. One solution may be to place a few flaps ("ears") in the lateral areas to incline the vapor jets and cause the liquid to move in the desired direction.

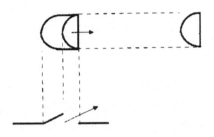

Figure 2.6. *Directive ear*

2.2. Drop in vapor pressure on crossing the plate

2.2.1. *Dry pressure drop (across the plate without liquid)*

According to Liebson *et al.* [LIE 57]:

$$\Delta P_S = \frac{1}{2}\rho_G \left[\frac{V_T}{C_o}\right]^2 \text{ or indeed } h_S = \frac{1}{2g}\left[\frac{V_T}{C_o}\right]^2 \frac{\rho_G}{\rho_L}$$

ΔP_S : dry pressure drop: Pa

ρ_G : density of the gaseous phase: $kg.m^{-3}$

V_T : velocity of the vapor on going through the holes: $m.s^{-1}$

h_S : dry load loss: m of column of liquid (m.C.L.)

According to Economopoulos [ECO 78]:

$$C_o = \left(0.836 + 0.273\frac{e_p}{d_T}\right) \times (0.674 + 0.717\phi)$$

e_p : thickness of the plate: m

ϕ : degree of piercing of the active surface of the plate:

$$\phi = \frac{\text{surface of the holes}}{\text{active surface}} \text{ ; in general: } 0.05 < \phi < 0.15$$

d_T : diameter of the holes: m

In general: $0.005 \text{ m} < d_T < 0.015 \text{ m}$

According to Economopoulos [ECO 78], it is also possible to use Hughmark and O'Connell's [OCO 57] formulae, or indeed those of Hunt *et al.* [HUN 55].

2.2.2. *Capillarity term in the pressure drop in the presence of liquid*

$$\Delta P_\sigma = \frac{\pi D_T \sigma}{\left(\pi d_T^2 / 4\right)} = \frac{4\sigma}{d_T}$$

σ : surface tension of the liquid: N.m^{-1}

Consider a load loss of:

$$h_\sigma = \frac{4\sigma}{g \rho_L d_T}$$

h_σ : capillary load loss: m.C.L.

The contribution of h_σ is generally negligible. Let us stress the fact that a pressure drop is measured in Pascals, whilst a load loss is measured in meters of liquid column. *A load is a height of liquid.*

2.2.3. *Pressure drop on crossing the liquid–vapor mixture*

Particularly in the jet regime, it is difficult to experimentally evaluate both the true height of the liquid–vapor mixture (LVM) on the plate and the true liquid fraction in volume of the LVM. We can get around this difficulty by directly expressing the pressure drop ΔP_L on crossing the LVM by:

$$h_m = (h_B + h_{LB})\beta \quad \text{with} \quad \Delta P_L = \rho_L g h_m$$

h_B : height of the dam at the outlet from the plate: m.C.L.

h_{LB} : height of liquid above the dam: m.C.L.

ρ_L : density of the liquid: $kg.m^{-3}$

β : correction coefficient (different from the volumetric fraction of liquid) according to Fair (reported by Economopoulos [ECO 78]):

$$\beta = 0.977 - 0.5075 \; F_A + 0.2292 \; F_A^2 - 0.035 \; F_A^3$$

F_A : kinetic parameter in relation to the active area on the plate:

$$F_A = V_A \sqrt{\rho_G} \; (kg^{0,5}.m^{-0,5}.s^{-1})$$

ρ_G : density of the vapor: $kg.m^{-3}$

V_A : velocity of the vapor in relation to the active area: $m.s^{-1}$

$$V_A = \frac{Q_G}{A_A}$$

The linear flowrate of liquid above a dam, after introduction of an empirical coefficient equal to 0.7, is given by Francis' formula:

$$\frac{Q_L}{L_B} = \frac{0.7 \times 2\sqrt{2g}}{3} h_{LB}^{3/2} \quad \text{and therefore} \quad h_{LB} = 0.61 \left(\frac{Q_L}{L_B}\right)^{2/3}$$

Q_L : frank liquid flowrate in terms of volume: $m^3.s^{-1}$

L_B : length of output dam: m

g : acceleration due to gravity: $9.81 \; m.s^{-2}$

2.2.4. *Pressure drop and height of the output dam*

The total pressure drop on the plate is:

$$\Delta P_T = \Delta P_S + \Delta P_\sigma + \Delta P_m = \Delta P_S + \Delta P_\sigma + \rho_L g \beta \left(h_B + h_{LB} \right)$$

and, after division by $\rho_L g$:

$$h_T = h_S + h_\sigma + \beta \left(h_B + h_{LB} \right)$$

If we take the total load loss h_T, we can deduce the height of the dam from it:

$$h_B = \frac{1}{\beta}(h_T - h_\sigma - h_S) - h_{LB}$$

In practice, we ignore the term h_σ.

A reasonable value of the total pressure drop of all the plates in the column must not be greater than 20% of the absolute operating pressure at the top of the column. In a vacuum, h_B may be negative. In this case, we need to increase the diameter of the column.

The common values for the height of the output dam are within the range (0.02 m – 0.08 m).

EXAMPLE 2.5.–

We agree that the vapor pressure drop across all of the plates represents 7% of the pressure at the head of the column. The column operates at 1.8 bar abs., and has 20 plates. Thus, we have the acceptable pressure drop per plate:

$$\Delta P_T = 10^5 \times 1.8 \times 0.07 / 20 = 630 \, \text{Pa}$$

However:

$$\rho_L = 800 \ \text{kg.m}^{-3}$$

The corresponding height of liquid is:

$$h_T = \frac{630}{800 \times 9.81} = 0.08 \text{ m.C.L.}$$

We have chosen 10 mm holes, and the plates are made of stainless steel. Hence:

$$e_p/d_T = 0.2$$

The degree of piercing (aperture) of the plate is 0.1, which corresponds to a triangular step p such that:

$$0.1 = \frac{\pi \times (0.01)^2}{2\sqrt{3}p^2}$$

Thus:

$$p = 0.03 \text{ m}$$

$$C_o = (0.836 + 0.273 \times 0.2) \times (0.674 + 0.717 \times 0.1)$$

$$C_o = 0.664$$

Dry load loss (expressed in height of clear liquid):

$$h_S = \frac{1}{2 \times 9.81} \left[\frac{1,55}{0.664 \times 0.1} \right]^2 \times \frac{1.25}{800} = 0.044 \text{ m.C.L.}$$

The length of the dam being 1.69 m, the height of liquid on the dam is (Francis' formula):

$$h_{LB} = 0.61 \left(\frac{0.02}{1.56} \right)^{2/3} = 0.0334 \text{ m.C.L.}$$

$$F_A = \frac{Q_G}{A_A}\sqrt{\rho_G} = \frac{11.66}{7.52}\sqrt{1.25} = 1.7335 \quad kg^{0,5}.m^{-0,5}.s^{-1}$$

$$\beta = 0.977 - 0.5075 \times 1.7335 + 0.2292 \times 1.7335^2 - 0.035 \times 1.7335^3$$

$$\beta = 0.603$$

The height of the dam is then:

$$h_B = \frac{1}{0.603}(0.08 - 0.044) - 0.0334 = 0.026 \quad m$$

2.3. Hydrodynamics of the plate

2.3.1. *General points on flooding*

The flooding of a column is expressed by an accumulation of liquid and an elevated pressure drop for the gaseous phase.

We distinguish two types of flooding:

– flooding by insufficient downcomer. This occurs when the liquid flowrate is too great to be channeled by the downcomer. The LVM accumulates in the downcomer and spills out onto the uppermost plate. We can remedy this situation by increasing the height of the downcomer – i.e. the space between the plates;

– flooding by entrainment of liquid droplets. In this case, the pressure drop becomes significant. To reduce the entrainment, we need to increase the spacing of the plates, which means that the drops have time to fall back onto the lower plate before they reach the upper one.

The result of the two types of flooding is the same and ultimately produces the accumulation of liquid in the spillway and a significant pressure drop on the side of the gas.

Note that a premature flooding may occur due to certain defects of the column – e.g. if the plate is not perfectly horizontal.

2.3.2. *Accumulation of liquid in the downcomer*

The height h_{LD} of liquid in the downcomer must balance:

1) the load loss at the "mouth" of the spillway – i.e. across the space existing between the deck and the next plate down. The liquid load corresponding to the outlet of the spillway is [FAI 63]:

$$h_{SD} = \frac{1}{2g}\left[\frac{Q_L}{L_B \times 0.6 \times h_{BO}}\right]^2 = 0.1525\left[\frac{Q_L}{A_{SD}}\right]^2$$

If the lower edge of the deck is rounded, 0.6 must be replaced by 1.

h_{SD}: height of liquid: m

h_{BO}: free height (height of the mouth) for the passage of the liquid under the deck: m

Q_L: flowrate of liquid: $m^3.s^{-1}$

A_{SD}: area available for the passage of the liquid between the lower plate and the underside of the apron: m^2

$$A_{SD} = h_{BO}L_B$$

According to Economopoulos [ECO 78], the surface of the mouth of the downcomer is equal, on average, to:

$$A_{SD} = 0,42A_D$$

A_D: horizontal section of the spillway: m^2

2) The height of liquid immediately at the outlet of the mouth of the downcomer on the lower plate:

$$h_B + h_{LB}$$

h_B: height of the dam of the downcomer: m

h_{LB}: height of liquid above the dam

3) The height of liquid h_T corresponds to the pressure drop of the gaseous phase on crossing the upper plate.

Thus, we are at the surface of the liquid–vapor mixture present in the downcomer.

NOTE.– To preserve a hydraulic joint which stops the vapor from climbing back into the downcomer, the free height for the input of liquid onto the plate must be less than 10–15 mm at the height of the outlet dam.

If the height of the outlet dam from the plate is slight (e.g. 2 cm) because the pressure drop across the plate is limited, the free height $= 1.56$ m is not sufficient to allow the liquid to pass, we need to use a lowered mouth (see Figure 2.2) or else a dam at the bottom of the deck. The height of that second dam would then be:

$$h_L + 0.02 \text{ m}$$

h_L : height of clear liquid on the plate: m

EXAMPLE 2.6.–

$$Q_L = 0.02 \text{ m}^3.\text{s}^{-1} \quad h_B = 0.026 \text{ m} \quad h_T = 0.08 \text{ m}$$

$$A_D = 0.2 \text{ m}^2 \quad h_{LB} = 0.033 \text{ m} \quad L_B = 1.56 \text{ m}$$

$$A_{SD} = 0.42 \times 0.2 = 0.084 \text{ m}^2$$

$$h_{BO} = 0.084 / 1.56 = 0.054 \text{ m}$$

As this height is greater than that of the dam, we need a lowered outlet (see Figure 2.2) with:

$$h_{SB} = h_{BO} - (h_B - 0.01) = 0.054 - (0.026 - 0.01) = 0.038 \text{ m}$$

$$h_{SD} = 0.1525 \left(\frac{0.02}{0.084} \right)^2 = 0.0086 \text{ m}$$

$$h_{LD} = 0.0086 + 0.026 + 0.033 + 0.08 = 0.148 \text{ m.C.L.}$$

Let us assume a value of 0.5 for the mean compactness of the LVM in the spillway. The compactness is the fraction of the volume occupied by the liquid. The height of the LVM is then:

$$0.148/0.5 = 0.30 \text{ m.C.L.}$$

This value is minimal for the spacing of the plates. Remember that, before, we had obtained a different minimal value:

$$S_{P\min} = 0.40 \text{ m.C.L.}\infty$$

It is this latter value which needs to be chosen for S_P.

NOTE.– In reality, the mean porosity $\overline{\varepsilon}$ of the LVM in the descent of the downcomer must be taken as equal to the arithmetic mean between zero (at the bottom) and ε_G (on the plate). The mean porosity is then $(1-\overline{\varepsilon})$.

$$\overline{\varepsilon} = \frac{1}{2}(0+\varepsilon_G) = \frac{\varepsilon_G}{2}$$

EXAMPLE 2.7.–

$$\varepsilon_G = 0.874 \text{ (see section 2.4.1)}$$

$$\overline{\varepsilon} = \frac{0.874}{2} = 0.437\#(1-0.5)$$

2.3.3. *Flooding by entrainment of drops of liquid*

Here, we define a kinetic parameter by:

$$F = \frac{Q_G}{A_S}\left(\frac{\rho_G}{\rho_L}\right)^{0,5}$$

A_S is the horizontal area of the volume separating the gaseous phase and the drops it has brought with it (entrained):

$$A_S = A_C - A_D$$

A_C and A_D are respectively the area of the section of the column and that of the downcomer.

Upon engorgement, Treybal [TRE 68] proposed the following form for the kinetic parameter F_E at flooding (obtained on the basis of the curves found by Fair [FAI 61]):

$$F_E = (0,0744\ S_P + 0,0117)Log_{10}X_M + 0,0304\ S_P + 0,0153$$

where:

$$X_M = \frac{Q_G}{A_S}\left(\frac{\rho_G}{\rho_L}\right)^{0,5}$$

The aperture of the plates is the ratio of the surface of all the holes to the active surface of the plate. When this aperture (which is represented by ϕ) is less than 0.1, we need to multiply F_E by the correction coefficient C_F:

$$C_F = \left(\frac{\sigma}{0.020}\right)^{0.2}\left(\frac{\phi}{0.1}\right)^{0.44}$$

The gas flowrate in terms of volume and on flooding is:

$$Q_{GE} = A_S F_E \left(\frac{\rho_L}{\rho_G}\right)^{0.5}$$

The approach to flooding is then:

$$E = \frac{Q_G}{Q_{GE}} = \frac{F}{F_E}$$

In practical terms, we must have:

$$0.6 < E < 0.8$$

EXAMPLE 2.8.–

$$\rho_L = 800 \text{ kg.m}^3 \qquad Q_L = 0.02 \text{ m}^3.\text{s}^{-1} \qquad A_C = 9 \text{ m}^2$$
$$\rho_G = 1.25 \text{ kg.m}^{-3} \qquad Q_G = 11.66 \text{ m}^3.\text{s}^{-1} \qquad A_D = 0.20 \text{ m}^2$$
$$\sigma = 0.020 \text{ N.m}^{-1} \qquad S_P = 0.4 \text{ m} \qquad \varphi = 0.1$$

$$X_M = \frac{11.66}{0.02}\left(\frac{1.25}{800}\right)^{0,5} = 23.04678$$

$$C_F = 1$$

$$F_E = (0.0744 \times 0.4 + 0.0117)\text{Log}_{10} 23.0468 + 0.0304 \times 0.4 + 0.0153$$

$$= 0.04146 \times 1.36236 + 0.01216 + 0.0153$$

$$F_E = 0.0839$$

$$Q_{GE} = 0.0839(9 - 0.20)\left(\frac{800}{1.25}\right)^{0.5} = 18.68$$

The approach to flooding is:

$$E = \frac{Q_G}{Q_{GE}} = \frac{11.66}{18.68} = 0.62$$

This value, which is relatively close to 0.7, is satisfactory.

NOTE.– [STI 78] proposes what he calls a maximum flowrate, characterized by:

$$F_{max} = 2.5\left[\phi^2 \sigma \ (\rho_L - \rho_G)g\right]^{0.25}$$

where:

$$V_{max} = F_{max}/\sqrt{\rho_G} \text{ and } Q_{G max} = V_{max} \times A_A$$

EXAMPLE 2.9.–

$$\varphi = 0.1 \qquad\qquad \rho_L = 800 \text{ kg.m}^3 \qquad g = 9.81 \text{ m.s}^{-2}$$

$$\sigma = 0.020 \text{ N.m}^{-1} \qquad \rho_G = 1.25 \text{ kg.m}^{-3} \qquad A_A = 7.38 \text{ m}^2$$

$$F_{max} = 2.5\left[0.2.10^{-3} \times 800 \times 9.81\right]^{0.25} = 2.80$$

$$V_{max} = 2.80/\sqrt{1.25} = 2.5 \text{ m.s}^{-1}$$

$$Q_{G\,max} = 2.5 \times 7.52 = 18.4 \text{ m}^3.\text{s}^{-1}$$

Note that this value is very close to Q_{GE}, according to Treybal, which is equal to $18.84 \text{ m}^3.\text{s}^{-1}$. For the calculation of the interfacial area, therefore, we use the method developed by Stichlmair and Mersmann [STI 78] but with an approach to engorgement in line with Treybal [TRE 68].

2.3.4. *Flowrate of entrained drops*

According to Fair [FAI 61], the degree of entrainment is defined by:

$$X_e = \frac{e}{L + e}$$

L: liquid mass flowrate: kg.s^{-1}

e: entrained liquid mass flowrate: kg.s^{-1}

The calculation of X_e involves the approach to flooding E and the flowrate parameter:

$$PD = \frac{Q_L}{Q_G}\sqrt{\frac{\rho_L}{\rho_G}}$$

Thus, we have the degree of entrainment:

$$X_e = \left[-(6.692 + 1.956\ E)\ PD^{(-0.132 + 0.654\ E)}\right]$$

EXAMPLE 2.10.–

$$E = 0.62 \qquad PD = 0.0434$$

$$X_e = \exp\left[-(6.692 + 1.956 \times 0.62)0.0434^{(-0.132+0.654\times0.62)}\right]$$

$$X_e = 0.035 = 3.5 \ \%$$

According to the rules of the discipline, the entrainment must not be greater than 8%.

2.3.5. *Heights of clear liquid and of the LVM*

According to Benett *et al.* [BEN 83]:

$$h_{LC} = \alpha_e\left[h_B + C\left(\frac{Q_L}{L_B\alpha_e}\right)^{0.67}\right]$$

$$\alpha_e = \exp\left[-12.55\left(V_A\left(\frac{\rho_G}{\rho_L}\right)^{0.5}\right)^{0.91}\right]$$

and $C = 0.50 + 0.438\exp(-137.8\ h_B)$

V_A : velocity of the gaseous phase expressed in relation to the active area: $m.s^{-1}$

h_B : height of the dam: m

L_B : length of the dam: m

The fraction of volume occupied by the gaseous phase is:

$$\varepsilon_G = (Q_G/Q_{GE})^{0.28} = E^{0.28} \ \text{[STI 78a]}$$

That is:

$$h_m = \frac{h_{LC}}{1-\varepsilon_G}$$

h_m : height of the LVM: m

EXAMPLE 2.11.–

$$h_B = 0.026 \text{ m} \qquad E = 0.62 \qquad \rho_G = 1.25 \text{ kg.m}^{-3}$$
$$Q_L = 0.02 \text{ m}^3.\text{s}^{-1} \qquad L_B = 1.56 \text{ m} \qquad \rho_L = 800 \text{ kg.m}^{-3}$$
$$Q_G = 11.66 \text{ m}^3.\text{s}^{-1} \qquad\qquad A_A = 7.52 \text{ m}^2$$

$$V_A = 11.6/7.52 = 1.55 \text{ m.s}^{-1}$$

$$\alpha_e = \exp\left[-12.55\left(1.55\left(\tfrac{1.25}{800}\right)^{0.5}\right)^{0.91}\right]$$

$$\alpha_e = 0.372$$

$$C = 0.50 + 0.438\exp(-137.8 \times 0.026)$$

$$C = 0.5121$$

$$h_{LC} = 0.372\left[0.026 + 0.5121\left(\frac{0.02}{0.372 \times 1.56}\right)^{0.67}\right]$$

$$h_{LC} = 0.030 \text{ m}$$

$$\varepsilon = 0.62^{0.28} = 0.875$$

$$h_m = 0.030 / (1 - 0.875) = 0.24 \text{ m}$$

2.3.6. *Beginning of weeping*

The correlation found by Lockett and Banik [LOC 84] can be written:

$$V_{Apl} = \phi \times 0.67 \left[\frac{gh_{cl}(\rho_L - \rho_G)}{\rho_G}\right]^{1/2}$$

The flexibility of a holed plate measures the relative decrease of the gaseous flowrate in order for weeping to occur:

$$S = \frac{V_A - V_{Apl}}{V_A}$$

EXAMPLE 2.12.–

$$g = 0.981 \text{ m.s}^{-2} \qquad h_{c\ell} = 0.03 \text{ m} \qquad \varphi = 0.1$$

$$\rho_L = 800 \text{ kg.m}^{-3} \qquad V_A = 1.55 \text{ m.s}^{-1} \qquad \rho_G = 1.25 \text{ kg.m}^{-3}$$

$$V_{Apl} = 0.1 \times 0.67 \left[\frac{9.81 \times 0.03 \times 800}{1.25} \right]^{1/2}$$

$$V_{Apl} = 0.92 \text{ m. s}^{-1}$$

If the plate works at a speed expressed in relation to the active area equal to 1,55 m.s^{-1}, its flexibility is:

$$S = \frac{1.55 - 0.92}{1.55} = 40 \text{ \%}$$

The flexibility of holed plates is less than that of bubble-cap plates, which can be up to or even greater than 70%. This property is exploited when, in the process or indeed along a column, the gaseous flowrate is highly variable, but this situation is essentially the only one for which bubble-cap plates are still used. [WUI 72] gives a calculation method for such plates, as does Bolles [BOL 56].

2.3.7. *Transition between the foam and jet regimes*

The gaseous flowrate calculated here corresponds to the jet regime established for all the holes of the plate.

According to Fell and Pinczewski [FEL 82], the velocity of the gas expressed in relation to the active area is:

$$V_{Atra} = \frac{2.75}{\sqrt{\rho_G}} \left[\frac{Q_L}{L_B} \sqrt{\rho_L} \right]^n \quad \text{where: } n = 0{,}91 \left(\frac{d_T}{\phi} \right)$$

ϕ : aperture of the plate

d_T : diameter of the holes: m

Q_L : flowrate of liquid: m^3.s^{-1}

L_B : length of the dam: m

ρ_L and ρ_G: densities of the liquid and the gas: kg.m^{-3}

EXAMPLE 2.13.–

$$V_A = 1.55 \ \text{m.s}^{-1}$$

$$V_{Atra} = \frac{2.75}{\sqrt{1.25}} \left[\frac{0.02}{1.69} \sqrt{800} \right]^n \quad n = \frac{0.91 \times 0.003}{0.1}$$

$$n = 0.0273 \quad V_{Atra} = 2.39 \ \text{m.s}^{-1}$$

However:

$$1.55 < 2.39$$

Thus, we are in the foam regime. Note that that correlation of Jeronimo *et al.* [JER 73] gives a smaller value for V_{Atra} but that value corresponds to a jet operation of only 70% of the holes.

2.4. Transfers of mass and heat

2.4.1. *Interfacial area [STI 78]*

If we know the operating regime (foam or jet), it is possible to find whether the LVM is presented in the form of gaseous bubbles in the liquid or indeed liquid droplets dispersed in the gaseous phase. The corresponding diameters are given by:

$$d_{bubble} = \left(\frac{6\sigma}{(\rho_L - \rho_G)g} \right)^{1/2} \quad d_{drop} = \frac{12\sigma\phi^2}{V_{AE}^2 \rho_0} \quad \text{where } V_{AE} = V_A/E$$

To evaluate the volume occupied by the dispersed phase, Stichlmair introduced the parameter $F = V_A \sqrt{\rho_G}$ and the ratio F/F_{max} where F_{max} corresponds to flooding by entrainment of liquid. We replace that ratio by the flooding approach E, and we write:

$$\varepsilon_G = E^{0.28} = 1 - \varepsilon_L \ [\text{STI 78a}]$$

ϵ_G and ϵ_L represent the volume fraction occupied by the gas and the liquid in the LVM. Expressed in relation to the volume of the MLV, the interfacial area is:

$$a = \frac{6\epsilon_D}{d}$$

ϵ_D is the fraction of volume occupied by the dispersed phase.

NOTE.– Certain authors advocate a correction term to be applied to a, but that term is not included in Stichlmair and Fair [STI 98]. However, a calculation for it is given in Appendix 1.

EXAMPLE 2.14.–

$$\sigma = 0.02 \text{ N.m}^{-1} \qquad \rho_L = 800 \text{ kg.m}^{-3} \qquad \rho_G = 1.25 \text{ kg.m}^{-3}$$
$$V_A = 1.55 \text{ m.s}^{-1} \qquad g = 9.81 \text{ m.s}^{-1} \qquad \varphi = 0.1$$
$$E = 0.62$$

$$d_b = \left(\frac{6 \times 0.02}{(800 - 1.25) \times 9.81} \right)^{1/2}$$

$$d_b = 0.0039 \text{ m}$$

$$d_g = \frac{12 \times 0.02 \times 0.01}{1.55^2 \times 1.25}$$

$$d_g = 0.0008 \text{ m}$$

$$\epsilon_G = 0.62^{0.28} = 0.874$$

$$\epsilon_L = 1 - 0.874 = 0.126$$

$$a_b = \frac{6 \times 0.874}{0.0039} = 1344 \text{ m}^{-1} \qquad a_g = \frac{6 \times 0.126}{8.10^{-4}} = 945 \text{ m}^{-1}$$

With these data, we have obtained the nature of the regime, which is a foam regime. We therefore need to employ the value a_b.

2.4.2. *Coefficient of mass transfer on the side of the gaseous phase*

According to Stichlmair [STI 78b]:

$$\beta_G^* = 2\left(\frac{D_G V_A}{\pi h_m \varepsilon_G}\right)^{1/2}$$

Accepting the validity of the ideal gas law, the total concentration is:

$$c_T = \frac{n}{V} = \frac{P}{RT}$$

Hence:

$$\beta_G = c_T \beta_G^* = \frac{2P}{RT}\left(\frac{D_G V_A}{\pi h_m \varepsilon_G}\right)^{1/2}$$

P : pressure: Pa

R : ideal gas constant: 8314 J.kmol^{-1}.K^{-1}

T : absolute temperature: K

V_A : velocity of the gaseous phase: m.s^{-1}

$$V_A = \frac{Q_G}{A_A}$$

A_A : active surface: m^2

h_m : height of the LVM (the "foam"): m

ε_G : fraction of the volume occupied by the gaseous phase

$$h_m = \frac{h_{LC}}{1-\varepsilon_G}$$

h_{LC} : height of clear liquid: m

EXAMPLE 2.15.–

$$P = 0.95.10^5 \text{ Pa} \quad h_{LC} = 0.030 \text{ m} \quad E = 0.62$$
$$T = 323 \text{ K} \quad V_A = 1.55 \text{ m.s}^{-1} \quad D_G = 0.3.10^{-4} \text{ m}^2.\text{s}^{-1}$$
$$\varepsilon_G = 0.874 \quad h_m = 0.24 \text{ m}$$

$$h_m = \frac{0.030}{1 - 0.874} = 0.24 \text{ m}$$

$$\beta_G = \frac{2 \times 0.95.10^5}{8314 \times 323} \left(\frac{0.3.10^{-4} \times 1.55}{\pi \times 0.24 \times 0.874} \right)^{1/2}$$

$$\beta_G = 0.594.10^{-3} \text{ kmol.m}^{-2}.\text{s}^{-1}$$

2.4.3. Coefficient of mass transfer on the side of the liquid phase

According to Stichlmair [STI 78b]:

$$\beta_L^* = 2 \left(\frac{D_L V_A}{\pi h_m \varepsilon_G} \right)^{0.5}$$

$$\beta_L = c_T \beta_T^*$$

c_T : total concentration of the liquid: kmol.m^{-3}

$$c_T = \frac{1}{\sum_i x_i \overline{V_i}} = \frac{\rho_L}{\sum_i x_i M_i}$$

$\overline{V_i}$: partial volume of the component i: m^3.kmol^{-1}

M_i : molar mass of the component i: kg.kmol^{-1}

x_i : molar fraction of the component i

EXAMPLE 2.16.–

$$c_T = 37 \text{ kmol.m}^{-3} \qquad D_L = 1.5.10^{-9} \text{ m}^2.s^{-1} \qquad V_A = 1.55 \text{ m.s}^{-1}$$

$$h_m = 0.24 \text{ m} \quad = 323 \text{ K} \qquad\qquad\qquad \varepsilon_G = 0.874$$

$$\beta_L^* = 2 \left(\frac{1.5.10^{-9} \times 1.55}{\pi \times 0.24 \times 0.874} \right)^{0,5}$$

$$\beta_L^* = 0.119.10^{-3} \text{ m.s}^{-1}$$

$$\beta_L = 37 \times 0.119.10^{-3} = 4.40.10^{-3} \text{ kmol.m}^{-2}.s^{-1}$$

2.4.4. *Overall transfer coefficient*

According to the simplified two-film theory (see section 4.2.2):

$$\frac{1}{K_L} = \frac{1}{\beta_L} + \frac{1}{m\beta_G} \quad \text{and} \quad \frac{1}{K_G} = \frac{m}{\beta_L} + \frac{1}{\beta_G} = \frac{m}{K_L}$$

m: slope of the equilibrium curve: $m = dy / dx$

EXAMPLE 2.17.–

$$m = 0.9 \ \beta_L = 4.40.10^{-3} \text{ kmol.m}^{-2}.s^{-1} \ \beta_G = 0.594.10^{-3} \text{ kmol.m}^{-2}.s^{-1}$$

$$\frac{1}{K_L} = \frac{10^3}{4.40} + \frac{10^3}{0.9 \times 0.594}$$

$$K_L = 0.476.10^{-3} \text{ kmol.m}^{-2}.s^{-1}$$

$$K_G = \frac{0.476.10^{-3}}{0.9} = 0.53.10^{-3} \text{ kmol.m}^{-2}.s^{-1}$$

2.4.5. *Arrangement of the mixture on the plate*

Consider the plate with index j, on which the liquid $(j-1)$ and the vapor $(j+1)$ arrive, having known compositions.

We shall suppose, to begin with, that the active area A_A has the shape of a rectangle whose sides are:

– ℓ_T: distance between the deck of the upstream downcomer and the dam of the downstream spillway – i.e. the length of the active area;

– L_B: length of the dam of the downcomers.

We shall also suppose that, along the distance ℓ_T, not only does the composition of the liquid evolve, but also the composition of the vapor coming from the lower plate is inconstant. On the other hand, along the dimension L_B, those two compositions are constant.

We shall begin by evaluating what we call the *local transfer* which takes place in the elementary volume:

$$d\omega = h_m L_B dl = h_m dA_A \ (A_A: \text{area active})$$

h_m : mean height of the LVM: m

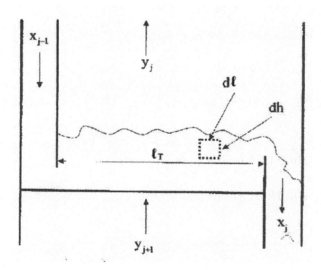

Figure 2.7. *Liquid–vapor contact on a plate*

2.4.6. *Local transfer (on a vertical)*

We shall make the hypothesis that the composition of the liquid is *constant on a vertical*. This results from the intense agitation existing in the LVM. The consequence of that is that the composition of the vapor at equilibrium with the liquid is also constant on a vertical. Let y_i^* represent that composition.

Consider the elementary volume (see Figure 2.7):

$$dV = L_B dh dl$$

The component with the index i transferred into that volume is:

$$K_{Gi}\,(y_i - y_i^*)adV$$

a: volumetric area for the transfer expressed in relation to the LVM: m^{-1}

The gaseous flowrate affected by this transfer is:

$$G\frac{dl}{l_T} \quad (G: kmol.\,s^{-1})$$

Thus, we have the balance relative to the gaseous phase:

$$K_{Gi}(y_i - y_i^*)adhL_B dl = \frac{dl}{l_T}Gdy_i$$

This means, after simplification by $d\ell$, that:

$$-\frac{dy_i}{y_i - y_i^*} = \frac{(L_B l_T)K_{Gi}adh}{G}$$

G: total gaseous flowrate: $kmol.s^{-1}$

Let us integrate over the thickness h_m of the LVM. We obtain the *local* composition of the gas on exiting the plate with index n (the plates are numbered from top to bottom of the column):

$$y_{j,i} = y_{j+1,i}e^{-N_{oGi}} + y_{j,i}^*\left(1 - e^{-N_{ogi}}\right) \qquad [2.1]$$

where:

$$N_{OGi} = \frac{L_B l_T K_{Gi} a h_m}{G} = \frac{A_A K_{Gi} a h_m}{G}$$

Note that the number of local transfer units is the same on the whole of the active surface of the plate if, though, we can accept that K_{Gi} does not vary.

2.4.7. Murphree efficiency

Let us set:

$$E_{MG} = 1 - e^{-N_{OG,i}}$$

Equation [2.1] becomes:

$$E_{MG} = \frac{y_{j,i} - y_{j+1,i}}{y_{j,i}^* - y_{j+1,i}}$$

This relation is the definition of the Murphree efficiency, which is valid over the whole of the active surface. On that surface, it must not be forgotten that $y_{n,i}^*$ is not constant, as we shall see.

EXAMPLE 2.18.–

$$N_{OG,i} = 2.57 \text{ and therefore } E_{MG} = 1 - e^{-2.57} = 0.923$$

2.4.8. Evolution of liquid on the plate

If the gas is impoverished, the liquid is enriched, and vice versa (owing to the conservation of material):

$$L dx_{j,i} + (y_{j,i} - y_{j+1,i}) dG = 0 \hspace{2cm} [2.2]$$

where:

$$dG = G \frac{dl}{l_T} \quad \text{and} \quad y_{j,i}^* = m_{j,i} x_{j,i} + x_{0,j,i}$$

However, we know that (see equation [2.1])

$$y_{j,i} - y_{j+1,i} = (1 - e^{-N_{OG,i}})(m_{j,i}x_{j,i} + x_{0,j,i} - y_{j+1,i})$$ [2.3]

By combining equations [2.2] and [2.3]:

$$\frac{L}{G} \frac{dx_{j,i}}{[1-\exp(N_{OG,i})](m_{j,i}x_{j,i}+x_{0,j,i}-y_{j+1,i})} = -\frac{dl}{l_T}$$ [2.4]

In general, $y_{j+1,i}$ is not constant on the plate $j+1$, and it is necessary to integrate equation [2.4] numerically. In the particular case that $y_{n-1,i}$ is constant, we would have:

$$\frac{L}{mG[1-\exp(-N_{OG,i})]} Ln\left[\frac{x_{j,i}+(x_{0,j,i}-y_{j+1,i})/m_{j,i}}{x_{j-1,i}+(x_{0,j,i}-y_{j+1,i})/m_{j,i}}\right] = -1$$

Let us set:

$$\lambda_{j,i} = \frac{mG}{L}[1-\exp(-N_{OG,i})]$$

Thus, we have the *transfer equation*:

$$x_{j,i} = x_{j-1,i}e^{-\lambda_i} + (y_{j+1,i} - x_{0,j,i})\left(\frac{1-e^{-\lambda_i}}{m_{j,i}}\right)$$

EXAMPLE 2.19.–

$x_{j-1,i} = 0.40$	$y_{j+1,i} = 0.65$
$A_A = 7.52$ m^2	$h_m = 0.24$ m
$a = 1344$ m^{-1}	$G = 0.5$ kmol.s^{-1}
$K_{Gi} = 0.530.10^{-3}$ kmol.m^{-2}.s^{-1}	$m_{j,i} = 0.9$
$L/G = 1.6$	$x_{o,j,i} = 0.25$

$$N_{OG,i} = (7.52 \times 0.530.10^{-3} \times 1344 \times 0.24)/0.5$$

$$N_{OG,i} = 2.57 \text{ and } \exp(-N_{OG,i}) = 0.077$$

$$\lambda_{j,i} = \frac{0.9}{1.6}(1 - 0.077) = 0.519$$

$$x_{j,i} = 0.40 \times 0.595 + (0.65 - 0.25)\left(\frac{1 - 0.595}{0.9}\right)$$

$$x_{j,i} = 0.418$$

2.4.9. *Mass balance on the side of the liquid*

When the liquid flowrate varies significantly between the inlet and the outlet of a plate, we need to divide the length ℓ_T of the liquid trajectory into elementary intervals.

Let $L_{j,q-1}$ and $L_{j,q}$ represent the flowrates at the inlet and outlet of the interval with index q and set:

$$w_{j,i,q}^L = \overline{L_{j,q}}(x_{j,i,q} - x_{j,i,q-1}) \qquad [2.5]$$

The relationship between these two molar fractions is a transfer equation of the same form as that which normally links $x_{j,i}$ and $x_{j+1,i}$. The number of intervals is such that, for each of them, we have:

$$\frac{W_{j,q}^L}{L_{j,q-1}} \leq 0,01 \quad \text{where:} \quad W_j^L = \sum_i w_{j,i,q}^L$$

We would then have:

$$\overline{L_{j,q}} = L_{j,q-1} + \frac{W_{j,q}^L}{2}$$

This value needs to be used in equation [2.5]. The calculation of each elementary interval, therefore, is iterative.

Finally:

$$L_{j,q} = L_{j,q-1} + W_{j,q}^L \quad \text{and} \quad W_j^L = \sum_q W_{j,q}^L$$

Hence:

$$L_j = L_{j-1} + W_j^L$$

This way of operating is coherent, if we calculate L_j on the basis of L_{j-1} or indeed L_{j-1} on the basis of L_j.

The calculation of the $x_{j,i}$ can be deduced from this by the relations:

$$L_j x_{j,i} - L_{j-1} x_{j-1,i} = w_{j,i}^L \text{ where: } w_{j,i}^L = \sum_q w_{j,i,q}^L$$

By summing on the i values, we can see that, automatically, we have:

$$\sum_i x_{j,i} = 1 \text{ if we had } \sum_i x_{j-1,i} = 1$$

and *vice versa*.

NOTE.–

This slice-wise calculation enables us to directly transpose, to the side of the liquid, the calculation of the differential extractors to express the influence of backmixing on the plate. For the axial dispersion coefficient, we could try a modified expression of that of the A.I.Ch.E journal.

$$KD_A^{0.5} = 0.6299 + 2.85V_A + 613.56\frac{Q_L}{L_B} + 29.97h_B$$

With the above data and $K = 3.10^3$:

$$D_A = 1.90.10^{-5} m^2.s^{-1}$$

2.4.10. *Material balance on the side of the vapor (unsteady regime)*

We have:

$$V_{j+1} - V_j = L_j - L_{j-1}$$

Thus, V_j as a function of V_{j+1}, and vice versa.

$$V_{j+1}y_{j+1,i} - V_j y_{j,i} = L_j x_{j,i} - L_{j-1}x_{j-1,i} + M_j \frac{d\overline{x}_{j,i}}{d\tau}$$

$y_{j,i}$ as a function of $y_{j+1,i}$, and vice versa.

The last term on the right-hand side refers to section 2.4.11.

By summing on i, we see that:

$$\text{If } \sum_i y_{j+1,i} = 1 \text{ then } \sum_i y_{j,i} = 1$$

and vice versa.

Logically, it would make sense to repeat the calculations for transfer on the side of the liquid with:

$$\overline{V} = \frac{1}{2}(V_{j+1} + V_j) \text{ and } \overline{L} = \frac{1}{2}(L_{j-1} + L_j)$$

2.4.11. *Mean composition of the liquid on a plate*

We shall discuss the most complex case, which is that of a plate where the level wavefront, representing a discontinuity in the liquid flowrate, progresses.

The level wavefront where the value of the flowrate L changes sharply moves at the speed

$$v_L = \frac{L^{(1)} / c_T}{L_B h_{LC}}$$

L_B : length of the dam (width of the active area): m

h_{LC} : height of clear liquid: m

$L^{(1)}$: liquid flowrate kmol.s^{-1}

c_T : total molar concentration: kmol.m^{-3}

Consider the band of active area with the breadth L_B and the horizontal thickness ℓ_F. This distance ℓ_F links the mouth of the upstream downcomer and the wavefront.

For $\ell \leq \ell_F$, the transfer equation is written thus (see section 2.4.8):

$$x_{j,i}(l) = b_{j,i} + \left(x_{j-1,i} - b_{j,i}\right)\exp\left(-\lambda_{j,i}\frac{l}{l_T}\right)$$

where:

$$b_{j,i} = \frac{y_{j+1,i} - x_{o,j,i}}{m_{j,i}} \qquad \lambda_{j,i} = \frac{m_{j,i}V_{j+1}}{L_{j-1}}(1 - e^{-N_{o,G,i}})$$

On the band of thickness ℓ_F, the mean value of $x_{j,i}$ is:

$$\overline{x_{j,1,F}} = \frac{1}{l_F}\int_0^{l_F} x_{j,i}dl = b_{j,i}\frac{l_T}{l_F\lambda_{j,i}}\left(x_{j-1,i} - b_{j,i}\right)\left[1 - \exp\left(-\lambda_{j,i}\frac{l_F}{l_T}\right)\right]$$

Similarly:

$$x_{j,i,(T-F)} = b_{j,i} + \left(x_{j,i,F} - b_{j,i}\right)\exp\left[-\lambda_{j,i}\frac{l_T - l_F}{l_T}\right]$$

$$\overline{x_{j,1,(T-F)}} = \frac{1}{(l_T l_F)}\int_{l_F}^{l_T} x_{j,i,(T-F)}dl$$
$$= b_{j,i} + \frac{l_T}{(l_T - l_F)}\left(x_{j,i,F} - b_{j,i}\right)\left\{1 - \exp\left[-\lambda_{j,i}\frac{(l_T - l_F)}{l_T}\right]\right\}$$

and, for the whole of the plate:

$$\overline{x_{j,i}} = \frac{1}{l_T}\left[l_F\overline{x_{j,i,F}} + (l_T - l_F)\overline{x_{j,i,(T-F)}^{(0)}}\right]$$

Having thus obtained an analytical expression for $x_{j,i}$, it is entirely possible to calculate its derivative with respect to time τ by writing:

$$\frac{d\overline{x_{j,i}}}{d\tau} = \frac{d\overline{x_{j,i}}}{dl_F} \cdot \frac{dl_F}{d\tau} = \frac{d\overline{x_{j,i}}}{dl_F} v_L$$

Indeed, as time τ elapses, the distance l_F increases proportionally to τ. In other words, the level wavefront moves across the plate at a constant velocity v_L.

EXAMPLE 2.20.–

$$x_{j-1,i} = 0.40 \qquad m_{j,i} = 0.9 \qquad y_{j+1,i} = 0.65$$

$$\lambda_{j,i} = 0.519 \qquad l_F = l_T \qquad x_{o,j,i} = 0.25$$

$$b_{j,i} = \frac{0.65 - 0.25}{0.9} = 0.4444....$$

$$\overline{x_{j,i}} = 0.4444.... + \frac{1}{0.519}(0.40 - 0.4444....)(1 - e^{-0.519})$$

$$\overline{x_{j,i}} = 0.40977$$

The mean value of $x_{j,i}$ is not too far from the arithmetic mean of $x_{j,i}$ and $x_{j-1,i}$.

Indeed, if we look again at section 2.4.8 where we found $x_{j,i} = 0.418$, we see that:

$$\frac{1}{2}(0.418 + 0.40) = 0.409 \# 0.40977$$

2.4.12. *Direction of calculation (bottom-up or top-down in the column)*

The above calculation was conducted with the supposition that we know the liquid and vapor arriving onto the plate. However, it is of greater interest to know:

– either the outgoing liquid and the incoming vapor, which defines the lower "inter-plate". In this case, we simply need to reverse the direction of the transfers, starting at the outlet of the liquid;

– or the incoming liquid and the outgoing vapor, which defines the upper inter-plate. Then, we need to make a hypothesis about the incoming vapor and, by calculation, deduce the outgoing vapor $y_{out}^{(0)}$ and we correct the incoming vapor $y_{in}^{(0)}$ by setting:

$$y_{in,i}^{(1)} = y_{in,i}^{(0)} \sqrt{\frac{y_{out,i}}{y_{out,i}^{(0)}}}$$

Having made these corrections, it is important not to forget to normalize the y_{in} by writing:

$$y_{in,i}^{N} = \frac{y_{in,i}}{\sum_{i} y_{in,i}} \quad (\text{"N" for "normalized"})$$

Thus, we can proceed along the column in one direction or the other by calculating the exchanges $\Delta w_{j,i}$ between the liquid and vapor on the plate j and pertaining to each component i.

2.4.13. *Equilibrium straight lines of the components*

Hereinafter, knowledge of what we call the "inter-plate j" means knowledge of the compositions $x_{j,i}$ and $y_{j+1,i}$. Furthermore, the asterisk means "at equilibrium with". We are able to calculate the inter-plates one after another, working upwards or downwards in a column (see section 2.4.12), but in order to do so we need to know the equilibrium straight lines of each component on the plate $j+1$.

Knowing the composition of the liquid L_j, we can calculate the composition of the vapor V_j^* at equilibrium with L_j at its boiling point. Similarly, if we know V_{j+1}, we can calculate the composition of L_{j+1}^* at the dew point of V_{j+1}. Thus, for each component, we know the two couples at equilibrium which characterize the inter-plate j situated beneath plate j.

$$(x_{j,i}, y_{j,i}^*) \text{ and } (x_{j+1,i}^*, y_{j+1,i})$$

From these two couples, we deduce the equilibrium line for the transfer taking place on the plate $j+1$ (see section 4.2.2 of the volume [DUR 16] of this set of books on thermodynamic), and leading to the inter-plate $j+1$. In the same way, we would calculate the inter-plate $j+2$ on the basis of $j+1$. In the calculation of the plate $j+1$, a refinement could consist of accepting, for both parameters of the equilibrium line, the arithmetic mean of the parameters of the lines for the inter-plates $j-1$ and j.

A computer with good precision (128 meaningful binary figures) is recommended for these calculations.

2.4.14. Heat transfer on the plate (see section 4.6 [DUR 16])

By replacing Q_I with its value in the expression of t_I, we obtain (see section 4.6.1 volume 1):

$$t_I = \frac{1}{A}(B_G t_G + B_L t_L + t_0)$$

the index I characterizes the liquid–gas interface with:

$$B_G = \frac{\alpha_G + \sum_i N_i C_{Gi}}{\alpha_G + \alpha_L} \quad B_L = \frac{\alpha_L - \sum_i N_i C_{Li}}{\alpha_G + \alpha_L}$$

$$A = 1 - \frac{\sum_i N_i (C_{Li} - C_{Gi})}{\alpha_G + \alpha_L} \qquad t_o = \frac{\sum_i N_i \Lambda_i}{\alpha_G + \alpha_L}$$

The meaning of the symbols is discussed in section 4.6.1 of [DUR 16].

Suppose the N_i are known and assimilated to their mean value on the plate. Indeed, taking account of the variations of N_i both across the height of the LVM and across the active surface would greatly complicate the calculations without significantly improving the accuracy. In an elementary volume dV, the heat exchanged is:

$$dq = \alpha_G (t_I - t_G) a dV = \alpha_G \left(\frac{B_G t_G}{A} + \frac{B_L t_L}{A} + \frac{t_O}{A} - t_G \right) a dV$$

We shall accept that the temperature of the liquid is constant on a vertical, and examine the variation in temperature of the gaseous phase.

$$dq = \alpha_G \left(1 - \frac{B_G}{A} \right) \left(t_L^* - t_G \right) a dV$$

where:

$$t_L^* = \frac{(B_L t_L + t_O)}{A - B_G}$$

Hence:

$$d\, t_G = \frac{\alpha_G}{G\, C_G} \left(1 - \frac{B_G}{A} \right) \left(t_L^* - t_G \right) a\, L_B l_T dh$$

Let us set:

$$\theta = \frac{\alpha_G}{G\ C_G} (1 - \frac{B_G}{A}) a A_A h_m$$

where:

L_B : length of the dam: m

ℓ_T : distance between the mouth and the dam: m

A_A : active area: m^2

$$A_a = L_B l_T$$

h_m : height of foam: m

a : interfacial area expressed in relation to the volume of foam: m^{-1}

θ : number of units of heat transfer on a vertical

On a vertical, we have:

$$\frac{dt_G}{t_G - t_L^*} = -\theta \frac{dh}{h_m} \quad \text{and therefore} \quad \text{Ln}\left[\frac{t_{G,j+1} - t_L^*}{t_{G,j} - t_L^*}\right] = e^{-\theta}$$

$$t_{Gj} - t_{G,j+1} = (t_L^* - t_{G,j+1})(1 - e^{-\theta})$$

As the liquid progresses toward the dam, its temperature t_{Lj} varies, because it takes heat from the gas. This gives us the heat balance:

$$LC_L dt_{Lj} = (t_{G,j+1} - t_{G,j})C_G dG \quad \text{where} \quad dG = \frac{dl}{l_T}$$

or indeed:

$$LC_L dt_L = -(t_L^* - t_{G,j+1})(1 - e^{-\theta})C_G \frac{dl}{l_T}G$$

However:

$$t_L^* = \frac{B_L}{A - B_G}(t_L + \frac{t_o}{B_L})$$

$$(t_L^* - t_{G,j+1}) = \frac{B_L}{A - B_G}\left[t_L + \frac{t_o}{B_L} - \frac{(A - B_G)t_{G,j+1}}{B_L}\right]$$

Let us set:

$$t_{G,j+1}^* = -\frac{t_o}{B_L} + \frac{(A - B_G)t_{G,j+1}}{B_L}$$

The heat balance is written:

$$LC_L dt_L = -\frac{B_L}{A-B_G}(t_L - t^*_{G,j+1})C_G \frac{dl}{l_T}G(1 - e^{-\theta})$$

Let us set:

$$\lambda_q = \frac{C_G G \; B_L(1-e^{-\theta})}{C_L L(A - B_G)}$$

We integrate:

$$t_{L,j} = t_{L,j-1} + (t^*_{G,j+1} - t_{L,j-1})(1-e^{-\lambda_q})$$

where, remember:

$$t^*_{G,j+1} = \frac{(A - B_G)t_{G,j+1}}{B_L} - \frac{\sum\limits_i N_i\Lambda_i}{(\alpha_G + \alpha_L)B_L}$$

2.4.15. *Calculation of the values used*

1) The mean value of the transfer flux densities is:

$$N_i = \frac{L(x_{j,i} - x_{j-1,i})}{A_A h_m a}$$

A_A : active area: m^2

h_m : height of LVM: m

a :   volumetric area of transfer: m^{-1}

2) The heat transfer coefficients are calculated in the same way as the material transfer coefficients:

$$\alpha_G = 2 \; C^*_G \rho_G \left(\frac{D_G V_A}{\pi h_m \varepsilon_m}\right)^{1/2} \qquad\qquad \alpha_L = 2 \; C^*_L \rho_L \left(\frac{D_L V_A}{\pi h_m \varepsilon_m}\right)^{1/2}$$

ρ_G and ρ_L : densities of the gas and the liquid

C_G^* and C_L^* : specific heat capacities of the gas and the liquid: $J.kg^{-1}.{}^{\circ}C^{-1}$

D_G and D_L : heat diffusivities of the gas and the liquid: $m^2.s^{-1}$

ε_m : porosity of the LVM (fraction of volume occupied by the gaseous phase)

h_m : height of the LVM: m

V_A : velocity of the gas across the active surface: $m.s^{-1}$

EXAMPLE.–

$x_{j,i}$ $= 0.418$	$A_A = 7.52$ m^2	ε_G $= 0.874$
$x_{j-1,i}$ $= 0.40$	$V_A = 1.55$ $m.s^{-1}$	h_m $= 0.24$ m
D_L $= 0.15.10^6$ $m^2.s^{-1}$	$L_B = 1.56$ m	D_G $= 20.10^{-6}$ $m^2.s^{-1}$
$C_L^*\rho_L = 2.5.10^6$ $J.m^{-3}.K^{-1}$	a $= 1344$ m^{-1}	$C_G^*\rho_G = 2.6.10^3$ $J.m^{-3}.K^{-1}$
C_L $= 5.10^4$ $J.kmol^{-1}.K^{-1}$	G $= 0.5$ $kmol.s^{-1}$	C_G $= 3.10^4$ $J.kmol^{-1}.K^{-1}$
$t_{L,j-1}$ $= 45^{\circ}C$	L $= 0.8$ $kmol.s^{-1}$	$t_{G,j+1}$ $= 50^{\circ}C$
$N_1 = 1,37.10^{-5}$ $kmol.m^{-2}.s^{-1}$		$N_2 = -0.2.10^{-5}$ $kmol.m^{-2}.s^{-1}$
Λ_1 $= 22.10^6$ $J.kmol^{-1}$		Λ_2 $= 37.10^6$ $J.kmol^{-1}$

$$N_1 = \frac{0.8(0.418-0.40)}{7.52 \times 0.24 \times 1344}$$

$$N_1 = 1.37.10^{-5} kmol.m^{-2}.s^{-1}$$

$$\alpha_G = 2600 \times 2 \left(\frac{20.10^{-6} \times 1.55}{\pi \times 0.24 \times 0.874} \right)^{1/2}$$

$$\alpha_G = 35.66 W.m^{-2}K^{-1}$$

$$\alpha_L = 2.5.10^6 \times 2 \left(\frac{0.15.10^{-6} \times 1.55}{\pi \times 0.24 \times 0.874} \right)^{1/2}$$

$$\alpha_L = 2970 \ \text{W.m}^{-2} \text{K}^{-1}$$

$$A = 1 - \frac{(1.37 - 0.2).10^5 (5-3)10^4}{2970 + 35.66} = 1 - 77.8.10^{-6}$$

$$A = 1$$

$$B_G = \frac{1}{3005.66} \left(35.66 + 3.10^4 (1.37 - 0.2)10^{-5} \right)$$

$$B_G = 0.011981$$

$$B_L = \frac{1}{3005.66} \left(2970 - 5.10^4 (1.37 - 0.2).10^{-5} \right)$$

$$B_L = 0.98794$$

$$t_o = \frac{1.37 \times 10^{-5} \times 22 \times 10^6 - 0.2 \times 10^{-5} \times 37 \times 10^6}{3005.66}$$

$$t_O = 0,07565°C$$

$$\theta = \frac{35.66}{0.5 \times 3.10^4} (1 - 0.011981) \ 1344 \times 7.52 \times 0.24$$

$$\theta = 5.697 \ \text{(see section 2.4.14)}$$

$$1 - e^{-\theta} = 0.9966$$

$$\lambda_q = \frac{3.10^4 \times 0.5 \times 0.98794 \times 0.9966}{5.10^4 \times 0.8(1 - 0.011981)}$$

$$\lambda_q = 0.3736 \qquad 1 - e^{-\lambda_q} = 0.3117$$

$$t^*_{G,j+1} = \frac{(1-0.011981)50}{0.98794} - \frac{0.07565}{0.98794}$$

$$t^*_{G,j+1} = 49.927°C$$

$$t_{L,j} = 45 + (49.927 - 45) \times 0.3117$$

$$t_{L,j} = 46.53°C$$

NOTE.– If the liquid flowrate varies noticeably, the path must be divided into elementary lengths, which will be the same as those used for the transfer of mass.

In addition, we should find that the vapor is slightly superheated and the liquid slightly supercooled. The opposite situation (supercooled vapor and superheated liquid) is, normally, unlikely to occur. In any case, though, whatever the situation, the equilibrium calculation will give a temperature (which we shall not use) and the composition of the two phases present.

More specifically, we shall calculate:

– the dew point of the vapor and the composition of that dew;

– the boiling point of the liquid and the composition of the bubbles.

2.4.16. Homogeneization of the vapor

We shall agree that there is homogenization over the section of the column for the vapor coming from a plate if:

$$\frac{D_C}{S_P} < 7$$

This limit may seem high, but it must be remembered that the direction of flow of the liquid over the active surface is reversed from one plate to the next.

Design and Performances of Liquid–Gas Packed Columns

3.1. General

3.1.1. *Principle of packed columns*

Packed columns are vertical cylinders filled with small solid bodies of varying shapes. Collectively, these solid bodies are known as the packing.

The gas is injected at the base of the column and recovered at the top. The liquid, for its part, is fed in to the top of the packing by various devices (feeders), the most effective of which ensure even distribution of the liquid throughout the section of the column.

3.1.2. *Important characteristic values of packed columns*

When the gas and liquid flowrates increase too much, a limit manifests itself, where the liquid has difficulty in descending, and where the drop in pressure of the gas on traversing the packing becomes excessive. This is known as flooding, and by studying it, we can link the diameter of the column to the desired flowrates.

The quantity of material exchanged by transfer between the gas and the liquid essentially depends on the nature of the packing and the height of the column. This height can be determined using the concepts of the material transfer coefficients and, sometimes, heights of transfer units.

3.1.3. *Usage and advantage of packed columns*

Packed columns are generally used with diameters of less than 0.3 m. In addition, their pressure drop on the side of the gas is less than that of a plate column with the same transfer performances.

However, these columns absolutely must be used when the diameter of the column is greater than 6 m, as is the case with the stripping of bromine from seawater, because an evenly-distributed flow of the liquid on too large, a plate would become difficult to deliver.

In a packed column, the liquid flows in a film and is not agitated by the gaseous phase as it is on a plate. The packing solution, therefore, is useful when dealing with foamy products.

To deal with a corrosive product, it is economical to use ceramic or plastic packing rather than manufacture plates of noble metal.

With an absolute pressure of less than 1 atm but greater than 0.2 atm, the pressure drop of the gas in a packed column is less than it would be in a plate column. Thus, packing is an appropriate solution when using moderate vacuums. For use in more powerful vacuums, manufacturers offer special, highly-porous packing.

Appendix 1 gives empirical relations regarding the vapor pressures for the absorption of hydrochloric gas and ammonia gas in water.

3.1.4. *Real height of a packed column*

Ultimately, the total height of packing H_T can be found by the following calculations:

– transfer of material across the effective area (effective height H_e);

– 25% increase of the effective height to take account of backmixing in the gaseous phase. We then obtain the useable height H_u;

– taking account of dead zones due to the distributors (feeder and recenterers) which neutralize a certain height of packing. This is the dead height, H_m.

3.2. Hydrodynamics of packed columns

3.2.1. *Physical significance of engorgement*

Upon flooding the liquid no longer flows in the column, because of viscous friction of the liquid against the packing but also the friction existing between the liquid and the gas.

The values we shall employ are:

A_c: section area of the column: m^2;

a: volumetric surface for flow of the liquid: m^{-1};

f_L, f_G: dimensionless friction factors (liquid and gas);

ρ_L, ρ_G: densities (liquid and gas): $kg.m^{-3}$;

Ω_L: volume of liquid present: m^3;

ε: porosity of the packing;

R: fraction of the void occupied by the liquid hold-up. Note that we are talking, here, about retention in relation to the void left by the packing rather than in relation to the volume of the column;

V_L, V_G: velocities in an empty bed of the liquid and the gas: $m.s^{-1}$;

U_G, U_L: real velocities of the gas and the liquid: $m.s^{-1}$.

The force of gravity drawing the liquid downwards is of the form:

$$F_1 = \rho_L \Omega_L g$$

The solid–liquid force of friction is expressed by:

$$F_2 = f_L A_c \frac{\rho_L V_L^2}{2}$$

The gas–liquid force of friction is expressed by:

$$F_3 = f_G A_c \frac{\rho_G V_G^2}{2}$$

At flooding there is, of course, an equilibrium relation, which is of the form:

$$F_1 = F_2 + F_3 \qquad [3.1]$$

Over a unitary height (1 meter), the volume of liquid Ω_L present in the column is:

$$\Omega_L = A_c \varepsilon R$$

Over the same height, the surface area over which the liquid flows is of the form:

$$A_e = A_c.a$$

The true velocities of the liquid and the gas are:

$$U_L = \frac{V_L}{\varepsilon R} \qquad U_G = \frac{V_G}{\varepsilon(1-R)}$$

By feeding these values back into the equilibrium equation [3.1], we find:

$$\left[\frac{\rho_L g \varepsilon^3}{\rho_G U_G^2 a} \right] \left[\frac{2R^3}{f_L} \right] = \frac{\rho_L V_L^2}{\rho_G V_G^2} + \frac{f_G}{f_L} \left[\frac{R}{1-R} \right]^2$$

However, f_L on the left-hand side of this equation can be replaced by $f' \mu_L^{0.2}$. This stems from a relation that is valid for the flow of a fluid in a pipe with $Re > 5000$ (see [MCA 63]).

Using the usual notations, the engorgement relation becomes:

$$\frac{1}{Y_E} \frac{2R^3}{f'} = X^2 + \frac{f_G}{f_L} \left[\frac{R}{1-R} \right]^2$$

Because the coefficients of friction and liquid retention R are functions of X, it made sense to try the following form:

$$\frac{1}{Y_E} = \alpha X^n + \beta X^m$$

The equation established by Tao [TAO 63] is:

$$\frac{1}{Y_E} = 30.7 \ X^{1.43} + 22 \ X^{0.40}$$

where:

$$X = \frac{L}{G}\sqrt{\frac{\rho_G}{\rho_L}} \text{ and } Y_E = \left[\frac{V_{GE}^2}{g}\right] \cdot \left[\frac{a}{\varepsilon^3}\right] \cdot \left[\frac{\rho_G}{\rho_L}\right] \cdot \left[\frac{\mu_L^{0.2}}{d_L}\right] \qquad [3.2]$$

d_L: density of liquid in relation to water (at 15°C): dimensionless;

L and G: mass flowrates of liquid and gas: kg.s^{-1};

μ_L: viscosity of the liquid: centipoises.

In reality, the expression of Y_E [3.2] is apt only when using 25 mm Raschig rings. For any other type of packing, the volumetric surface area a_T of the packing must be corrected by a form coefficient C_F. In addition, in this correlation, we take the total volumetric surface of the packing a_T in relation to the unit volume of the column.

The expression of Y_E then becomes:

$$Y_E = \left[\frac{V_{GE}^2}{g}\right] \cdot \left[\frac{C_F a_T}{\varepsilon^3}\right] \cdot \left[\frac{\rho_G}{\rho_L}\right] \cdot \left[\frac{\mu_L^{0,2}}{d_L}\right]$$

Table 3.1 gives the form coefficient C_F for various types of packing (see [PRA 69]).

Nature of the packing	Form coefficient C_F
Ceramic Raschig rings	$(30/d_N)^{0,26}$
Steel Raschig rings	1.6
Pall rings	0.5
Berl saddles	0.6
Intalox saddles	$(4,4/d_N)^{0,29}$

Table 3.1. *Form coefficient for flooding*

d_N: nominal dimension of the packing: mm

The table in Appendix J gives the values of a_T and ε for the different types of packing.

3.2.2. *Rate of engorgement (approach to engorgement)*

We can distinguish between two rates of engorgement:

1) Case of engineering:

We look for the diameter D_c of the column and impose L and G, and therefore L/G, and thus X. We also impose an approach to engorgement E_I:

$$E_I = \frac{V_G}{V_{GE}} = \frac{G}{G_E} = \frac{L}{L_E}$$

The diameter of the column is then:

$$D_c = \sqrt{\frac{G}{E_I V_{GE} \rho_G \pi / 4}}$$

It is commonplace to adopt:

$$0.7 < E_I < 0.8$$

In this zone, the liquid retention and the pressure drop start to increase rapidly with V_G. This is the loading zone. It is also the zone where the intensity of the material transfer is maximal.

2) Case of an existing column:

Here, ϕ and L are fixed, and we are looking to find how much we can increase V_G. This represents, for example, the flowrate of gas in an absorption column or distillation column. That flowrate is initially equal to $G^{(0)}$. We calculate:

$$X^{(0)} = \frac{L}{G} \sqrt{\frac{\rho_G}{\rho_L}} \rightarrow Y_E^{(0)} \rightarrow V_{GE}^{(0)}$$

$$X^{(1)} = X^{(0)} \left(V_G / V_{GE}^{(0)} \right) \rightarrow Y_E^{(1)} \rightarrow V_{GE}^{(1)}$$

$$X^{(2)} = X^{(0)} \left(V_G / V_{GE}^{(1)} \right) \rightarrow Y_E^{(2)} \rightarrow V_{GE}^{(2)}$$

$$X^{(3)} = X^{(0)} \left(V_G / V_{GE}^{(2)} \right) \rightarrow Y_E^{(3)} \rightarrow V_{GE}^{(3)}$$

In general, three iterations are sufficient, and we obtain:

$$E_I = V_G / V_{GE}^{(3)}$$

Predictably, if L remains constant, the increase in gas flowrate facilitated will be greater than in the case of engineering.

3.2.3. *Pressure drop*

The pressure drop of the gas is given by:

$$\Delta P = \frac{(7762 + 8762 \times X) Y \times H_T}{1 - Y(41 \times X + 0.6)}$$

H_T: total height of packing: m

ΔP: pressure drop: Pa

X and Y are calculated with the true values:

$$X = \frac{L}{G} \sqrt{\frac{\rho_G}{\rho_L}}; \qquad Y = \left[\frac{V_G^2}{g} \right] \cdot \left[\frac{C_F a_T}{\varepsilon^3} \right] \cdot \left[\frac{\rho_G}{\rho_L} \right] \cdot \left[\frac{\mu_L^{0,2}}{d_N} \right]$$

EXAMPLE 3.1.–

Ceramic Raschig rings:

$G = 0,83 \text{ kg.s}^{-1}$ $\qquad d_N = 0,05 \text{ m}$ $\qquad L = 1,95 \text{ kg.s}^{-1}$

$\rho_G = 1 \text{ kg.m}^{-3}$ $\qquad a_T = 130 \text{ m}^{-1}$ $\qquad \rho_L = 1000 \text{ kg.m}^{-3}$

$$H_T = 2 \text{ m} \qquad \varepsilon = 0.77 \qquad \mu_L = 1 \text{ centipoise}$$

$$C_F = (30/50)^{0.26} = 0.88$$

$$X = \frac{1.95}{0.83}\sqrt{\frac{1}{1000}} = 0.074$$

$$\frac{1}{Y_E} = 30.7(0.074)^{1.43} + 22(0.074)^{0.40}$$

$$Y_E = 0.1176$$

$$0.1176 = \frac{V_{GE}^2 \times 130 \times 0.88 \times 1 \times 1}{9.81 \times (0.77)^3 \times 1000 \times 1} = V_{GE}^2 \times 0.0257$$

$$V_{GE} = 2.14 \text{ m.s}^{-1}$$

Let us choose (case of engineering):

$$E_I = 0.7$$

Thus:

$$V_G = 2.14 \times 0.7 = 1.5 \text{ m.s}^{-1}$$

$$D_c = \sqrt{\frac{0.83}{0.7 \times 2.14 \times 1 \times 0.785}} = 0.84 \text{ m}$$

$$Y = 0.1176 \times \left[\frac{1.5}{2.14}\right]^2 = 0.058$$

$$\frac{\Delta P}{H} = \frac{(7762 + 8762 \times 0.074)0.058}{1 - 0.058(41 \times 0.074 + 0.6)} = 617 \text{ Pa.m}^{-1}$$

$$\Delta P = 617 \times 2 = 1234 \text{ Pa} = 0.012 \text{ bar}$$

Suppose that this column exists. By how much do we need to increase V_G to achieve engorgement with a constant liquid flowrate?

	X_E	Y_E	V_{GE}
0	0.074	0.1176	2.14
1	0.052	0.14	2.33
2	0.048	0.144	2.36
3	0.047	0.146	2.38

Table 3.2. *Iterative calculation*

The answer to this question is given in Table 3.2.

$$V_{GE} = 2.38 \text{ m.s}^{-1}$$

Thus:

$$V_{GE}/V_G = 2.38/1.5 = 1.59$$

Therefore, we need to increase V_G by 59%. In case 1 (L/G = const.), the relative increase in V_G would be only $1/0.7 = 1.43$, which is 43%.

3.2.4. *Liquid retention*

When studying chemical reactions, it may be useful to be aware of the hold-up. The expression given by Jesser and Elgin [JES 43], in SI units, is written:

$$\varphi = 1.0357 \times \left(\frac{L}{d_p}\right)^{0.6} \times \mu_L^{0.1} \times \rho_L^{-0.78} \times \left(\frac{0.072}{\sigma}\right)^n$$

with:

$$n = 0.465 - 0.0108 \, L$$

L: mass flux density of the liquid in an empty bed: $\text{kg.m}^{-2}.\text{s}^{-1}$

φ: volume of liquid expressed in relation to the unit volume of the column

ρ_L: density of the liquid: $kg.m^{-3}$

σ: surface tension of the liquid: $N.m^{-1}$

d_p: diameter of the sphere with the same surface area as the packing: m

$$d_p = Kd_N$$

d_N: nominal size of packing: m

The coefficient K is given in Table 3.3.

Type of packing	K
Raschig rings (ceramic)	0.85
Berl saddles (ceramic)	0.78
Pall rings (steel)	0.55
Pall rings (ceramic)	0.80
Intalox saddles (ceramic)	0.73

Table 3.3. *Jesser and Elgin's coefficient*

Details of how to calculate K are given in [WUI 72].

EXAMPLE 3.2.–

$$L = 3.52 \ kg.m^{-2}.s^{-1} \quad \rho_L = 800 \ kg.m^{-3} \quad d_N = 0.025 \ m$$

$$\mu_L = 1 \ cp = 10^{-3} \ Pa.s \quad \sigma = 0.02 \ N.m^{-1} \quad K = 0.85 \ (Raschig \ rings)$$

$$n = 0.465 - 0.0108 \times 3.52 = 0.428$$

$$\phi = 1.0357 \times \left(\frac{3.52}{0.025 \times 0.85}\right)^{0.6} \times (10^{-3})^{0.1} \times 800^{-0.78} \times \left(\frac{0.072}{0.020}\right)^{0.428}$$

$$\phi = 0.105$$

3.3. Effective height and useable height for the transfer

IMPORTANT NOTE.–

The calculation methods presented below are valid only for an *approach to engorgement equal to 0.7 (70%)* and with L/G = const.

3.3.1. *Methodology*

Suppose we have chosen a particular type of packing and calculated the diameter of the column. We then determine:

– the effective area of contact between the gaseous phase and liquid phase;

– the transfer coefficients of material and possibly of heat;

– the useable height that actually contribute to the transfer obtained by increasing the effective height by 25%;

– the dead height, relating to the imperfect distribution of the liquid. The height of packing is the sum of these two heights.

3.3.2. *Effective area for the transfer*

This area can be expressed as follows:

$$a_e = (ua_T - a_I)F_M$$

a_e: effective volumetric area expressed in relation to the volume of the column: m^{-1}

a_T: total volumetric area of the packing, also in relation to the volume of the column: m^{-1}

u: degree of wetting (fraction of the total area that is truly wetted)

a_I: volumetric area wetted but ineffective (stagnant recesses): m^{-1}

The total area a_T is a given value, supplied by the manufacturer (see Appendix 2). However, let us consider several types packing of the same nature but different sizes, and characterize that size by a similarity ratio k, which is the ratio between the size of the packing in question and that of the reference packing. In this case, the number of elements contained in the unit volume varies with k^{-3}, and the surface of each element with k^2. The total volumetric area a_T therefore varies with k^{-1}.

The meaning of the use rate u is as follows: if we place a piece of sheet metal underneath a flowing faucet, there is a very high probability that only one of the two faces of that metal will be irrigated. Furthermore, that wetted surface will probably actually be wetted only partly. This being the case, the use rate u of the surface of the sheet would be less than 0.5, and equal to, say, 0.3. Note that this factor depends only on the shape of the packing; not its dimension.

The surface a_I is the surface area of the recesses where there is stagnant liquid that is not renewed and, consequently, does not contribute to the material transfer.

A meniscus situated at the edge between two planes has a surface area which depends only on the length of that edge. The area of each meniscus therefore varies proportionally to k.

The number of menisci contained in 1 m^3 of packing varies with the number of packing elements contained in 1 m^3, i.e. with k^{-3}.

Ultimately, a_I varies with k^{-2}.

Here, we propose to use the values of u and a_I that are given in Table 3.4.

Nature of the packing	u	a_I
Raschig rings	0.29	20
Pall rings	0.34	26
Berl saddles	0.33	35
Intalox saddles	0.34	35

Table 3.4. Coefficients for efficient area

In this table, a_I is given for *25 mm elements*. For elements whose size d_N is different, we take $k = d_N/25$, where the nominal dimension d_N is measured in mm. These values correspond to operation with a degree of flooding equal to 70% (where the height of the transfer unit is minimal).

Let us examine the way in which the value $(ua_T - a_I)$ varies with the size of the packing in the case of Raschig rings.

K	a_T	a_I	u	$ua_T - a_I$
1	200	20	0.29	38
½	370	80	0.29	27
2	95	5	0.29	22

Table 3.5. *Effective area of transfer*

Note that the effective area of the packing in question is *maximum* for k = 1. This somewhat-surprising result concurs with the experimental results obtained by certain authors, who affirm that Raschig rings of 25 mm are the most plus effective. However, larger rings accept a higher flowrate of liquid at a constant degree of flooding. Thus, we can categorically state that small rings are not necessarily more effective than large ones.

Certain manufacturers offer high-performance packings made of vertical parallel plates that are corrugated to a greater or lesser degree, and of varying distances apart. Thus, the liquid flows over the two faces and there are no recesses of stagnant liquid. A limit to how close the plates can come to one another is the need for the gas to have sufficient room to maintain the pressure drop at a value lower than those of conventional packings.

Now let us examine the influence of the nature of the substances involved: in order for the column to really be irrigated, obviously, the liquid must be able to wet the packing. For this purpose, it may be favorable to look for a rough surface (rather than a polished one). However, if the liquid does not wet the packing and, in an extreme case, rolls on its surface in the form of drops, it is entirely possible that the transfer between the gas and the liquid will be acceptable, on condition that the drops are small enough. Therefore, the wettability of the packing is not necessarily a determining criterion, and the proof is that packing made of Teflon or graphite, which are hydrophobic, yield satisfactory results with an aqueous liquid phase.

A low surface tension encourages wetting of the packing and, if it is hydrophobic, favors the presence of small droplets, which therefore have a large surface area per unit volume. For this reason, we introduce an empirical factor F_m which multiplies the effective area:

$$F_M = \left[\frac{0.073}{\sigma} \right]^{0.25}$$

σ: surface tension of the liquid: $N.m^{-1}$

(Remember that the surface tension of pure water is $0,073\ N.m^{-1}$).

Of course, this expression of F_M is valid only in the absence of a tensioactive agent. Indeed, such an agent, whilst it does indeed increase the wetting of the packing, presents the major drawback of accumulating on the surface of the liquid and thus forming a barrier to the transfer of material between the liquid and the gas.

The effective area is often only around 10–20% of the total volumetric area, which is unsurprising. Indeed, tests have shown that half the liquid flows over 5% of the surface a_T, and that 50% of that surface is not in contact with any liquid at all.

The effective area can be measured directly by the sodium sulfite oxidation method.

3.3.3. *Partial transfer coefficient on the side of the gas*

The coefficient on the side of the gas is expressed as follows:

$$\beta_G = G.\frac{j_D}{Sc_G^{2/3}}$$

where:

$$j_D = 0,76\frac{F_R}{Re^{0,36}}$$

β_G: material transfer coefficient on the side of the gas: $kmol.m^{-2}.s^{-1}$

G: total molar flux density (in an empty bed): $kmol.m^{-2}.s^{-1}$

Sc_G: Schmidt number on the side of the gas:

$$Sc_G = \frac{\mu_G}{\rho_G D_G}$$

Re_G: Reynolds number on the side of the gas: $Re_G = \dfrac{6GM_G}{\mu_G a_T}$

M_G: mean molar mass of the gas: kg.kmol^{-1}

F_R: renewal factor: dimensionless

The factor F_R expresses the aptitude of the packing to divide the gaseous flow and to create and renew the gaseous film, and so to favor transfer.

Thus, when a fluid flows over a crest which divides that fluid to form two films, the local transfer coefficient (or heat or material) is considerable (if not infinite) along the edge. On the other hand, the surface involved is small, except when the packing is entirely composed of wires or vertical plates not in contact with each other, along which the liquid flows.

We use the term "renewal perimeter" to speak of the total length of the edges contained in the unitary volume of the packing.

Thus, the renewal perimeter of a Raschig ring is equal to the perimeter of one of the two circles which form its ends. Along the second circle, the film created by the first disappears, and this is true on the statistical level: half of the existing edges serve to create exchange films, while the other half eliminate those films.

The renewal perimeter P_R for usual packings for the nominal dimension of 25 mm is shown in Table 3.6.

Nature of the packing	P_{R25}
Raschig rings	$7.8 . 10^{-2}$
Pall rings	$30 . 10^{-2}$
Berl saddles	$10 . 10^{-2}$
Intalox saddles	$14 . 10^{-2}$

Table 3.6. *Renewal perimeter*

For the above packings, we shall accept that in the expression of the renewal factor F_R, the renewal perimeter P_R comes into play with the power 2/3.

When the nominal size d_N of the packing varies, the perimeter of each element varies in accordance with d_N. The number of elements contained in the unitary volume varies with d_N^{-3}. Therefore, the renewal perimeter varies with d_N^{-2}.

The result of the above is that the renewal factor is of the form:

$$F_R = K \left[\frac{P_{R25}}{d_N^2} \right]^{2/3}$$

In addition, with Raschig rings of 0.025 m in diameter, we shall agree that the factor F_R is equal to the unit, so:

$$F_R = 0.0415 \left[\frac{P_{R25}}{d_N^2} \right]^{2/3}$$

d_N: nominal size of the packing: m

P_{R25}: renewal perimeter of the element of nominal size 0.025 m: m

3.3.4. *Partial transfer coefficient on the side of the liquid*

We shall use a modification of the expression developed by Van Krevelen and Hoftijzer [VAN 47].

$$\beta_L = c_T \frac{D_L}{\delta_F} \times 0.012 \, Re_L^{0.66} \, Sc_L^{0.33}$$

β_L: material transfer coefficient on the side of the liquid: $kmol.m^{-2}.s^{-1}$

c_T: total concentration of the liquid in terms of all its components (including the solvent): $kmol.m^{-3}$

D_L: diffusivity of material: $m^2.s^{-1}$

δ_F: thickness of the limiting film: m

$$\delta_F = \left[\frac{\mu_L^2}{\rho_L^2 g} \right]^{1/3} \qquad\qquad [3.3]$$

μ_L: viscosity of the liquid: Pa.s

ρ_L: density of the liquid: kg.m^{-3}

g: acceleration due to gravity: m.s^{-2}

$$g = 9.81\,\text{m.s}^{-2}$$

Re_L: Reynolds number on the side of the liquid:

$$Re_L = \frac{LM_2}{\mu_L a_e}$$

LM_2: mass flux density (in an empty bed) of the liquid: kg.m^{-2}.s^{-1}

Sc_L: Schmidt number for the liquid solution:

$$Sc_L = \frac{\mu_L}{\rho_L D_L}$$

3.3.5. Heat transfer coefficients

Using the Chilton–Colburn analogy and examining the dimensions, we obtain the expressions below:

$$\alpha_L = 0.012 \frac{\lambda_L}{\delta_F} Re_L^{0.66} Pr_L^{0.33}$$

The thickness δ_F is the same as for material transfer (equation [3.3]).

$$\alpha_G = \frac{j_T G C_{pG}}{Pr_G^{2/3}}$$

The factor j_T is equal to j_D.

G: mass flux density (in an empty bed) for the gas: kmol.m^{-2}.s^{-1}

C_{pG}: molar specific heat capacity of the gas: J.kmol^{-1}.°C^{-1}

Pr_L and Pr_G: Prandtl numbers on the side of the liquid and of the gas:

$$Pr_L = \frac{C_{pL}\mu_L}{\lambda_L} \text{ and } Pr_G = \frac{C_{pG}\mu_G}{\lambda_G}$$

M_L and M_G: mean molar masses of the liquid and the gas: $kg.kmol^{-1}$

α_L and α_G: heat transfer coefficients on the side of the liquid and the gas: $W.m^{-2}.°C^{-1}$

λ_G and λ_L: heat conductivities of the gas and the liquid: $W.m^{-1}.°C^{-1}$

The other parameters have the same meaning as before.

3.3.6. *Height of transfer unit and useable height*

We have seen (in section 3.2.2):

$$\frac{1}{K_G} = \frac{1}{\beta_G} + \frac{m}{\beta_L} \tag{3.4}$$

m: slope dy/dx of the equilibrium curve

β_G, β_L: partial transfer coefficients in the gas and the liquid: $kmol.m^{-2}.s^{-1}$

x, y: molar fractions of the solute in the liquid and the gas

K_G: global transfer coefficient expressed in relation to the gas: $kmol.m^{-2}.s^{-1}$

Let us write a material balance for a segment of the column of height dH:

$$Gdy = K_G(y^* - y)a_e dH$$

Remember that L and G are the molar flux densities of liquid and gas in relation to the area section of the column ($kmol.m^{-2}.s^{-1}$).

a_e: effective volumetric area of transfer: m^{-1}

y^*: molar fraction in the gas at equilibrium with the liquid whose molar fraction is x

The *effective height* of the column is:

$$H_e = \int_0^H dH = \frac{G}{K_G a_e} \int_0^H \frac{dy}{(y^* - y)} = H_{OG} \times NTU$$

H_{OG}: height of a transfer unit to the side of the gas: m

NTU: number of transfer units

Multiply, by $\dfrac{G}{a_e}$, both sides of the expression of $1/K_G$ [3.4]:

$$\frac{G}{K_G a_e} = \frac{G}{\beta_G a_e} + \left[\frac{mG}{L}\right]\frac{L}{\beta_L a_e}$$

The partial heights of transfer unit are defined by:

$$H_G = \frac{G}{\beta_G a_e} \text{ and } H_L = \frac{L}{\beta_L a_e}$$

The global height of the transfer unit on the side of the gas is then:

$$H_{OG} = H_G + \frac{mG}{L} H_L$$

Finally, the useable height H_u is:

$$H_u = 1.25 H_e = 1.25 H_{OG} NUT$$

3.3.7. *Directly calculating the partial heights of transfer units*

The expressions given in this section correspond to the case where the distribution of the liquid is perfect and there is no wall effect.

1) On the side of the gas:

$$H_G = \frac{1.33}{a_e}.Re_G^{0.36}.Sc_G^{0.66}.F_R^{-1}$$

In this expression, we see the following values play a part (already defined in the discussion of the coefficients β_G and β_L):

Re_G: Reynolds number on the side of the gas:

$$Re_G = \frac{6GM_G}{\mu_G a_T}$$

M_G: mean molar mass of the gas: $kg.kmol^{-1}$

G: molar flux density (in an empty bed) of the gas: $kmol.m^{-2}.s^{-1}$

μ_G: viscosity of the gaseous phase: Pa.s

a_T: total volumetric area of the packing: m^{-1}

Sc_G: Schmidt number on the side of the gas

$$Sc_G = \frac{\mu_G}{\rho_G D_G}$$

ρ_G: density of the gas: $kg.m^{-3}$

D_G: diffusivity in the gas of the transferred component: $m^2.s^{-1}$

H_G: partial height of transfer unit on the side of the gas: m

F_R: renewal factor (as explained for the calculation of β_G)

This expression is consistent with that which was given for β_G.

2) On the side of the liquid:

$$H_L = 21\delta_F . Re_L^{0.34} . Sc_L^{0.67}$$

H_L: partial height of transfer unit on the side of the liquid: m

The expressions of δ_F, Re_L and Sc_L are the same as those which were used for the calculation of β_L.

This expression of H_L is consistent with that given for β_L.

EXAMPLE 3.3.–

Packed column using 50 mm Raschig rings

V_G	$= 1.5 \text{ m.s}^{-1}$	W_L	$= 1.95 \text{ kg.s}^{-1}$	σ	$= 0.030 \text{ N.m}^{-1}$
μ_G	$= 20.10^{-6} \text{ Pa.s}$	μ_L	$= 10^{-3} \text{ Pa.s}$	D_c	$= 0.84 \text{ m}$
ρ_G	$= 1 \text{ kg.m}^{-3}$	ρ_L	$= 1000 \text{ kg.m}^{-3}$	m	$= 1.29$
M_G	$= 32 \text{ kg.kmol}^{-1}$	M_L	$= 18 \text{ kg.kmol}^{-1}$	u	$= 0.29$
D_G	$= 0.152.10^{-4} \text{ m}^2.\text{s}^{-1}$	D_L	$= 0.20.10^{-8} \text{ m}^2.\text{s}^{-1}$	a_T	$= 95 \text{ m}^{-1}$
c_{TG}	$= 0.03125 \text{ kmol.m}^{-3}$	c_{TL}	$= 55 \text{ kmol.m}^{-3}$	a_I	$= 5 \text{ m}^{-1}$
V_G	$= 1.5 \text{ m.s}^{-1}$	g	$= 9.81 \text{ m.s}^{-2}$	W_L	$= 1.95 \text{ kg.s}^{-1}$

1) Effective area:

$$F_M = \left[\frac{0.073}{0.030}\right]^{0.25} = 1.25$$

$$a_e = (0.29 \times 95 - 5) \times 1.25 = 28.19 \text{ m}^{-1}$$

2) Coefficient on the side of the gas:

$$F_R = 0.0415\left[\frac{30.10^{-2}}{(50.10^{-3})^2}\right]^{2/3} = 1.0093$$

$$Re_G = \frac{6 \times (1,5 \times 1)}{20.10^{-6} \times 95} = 4736$$

$$j_D = 0.76.\frac{1.0093}{4736^{0.36}} = 0.0365$$

$$Sc_G = \frac{20.10^{-6}}{1 \times 0.152.10^{-4}} = 1.32$$

$$\beta_G = 1.5 \times 0.03125 \times \frac{0.0365}{1.32^{2/3}}$$

$$\beta_G = 0.00142 \text{ kmol.m}^{-2}.s^{-1}$$

3) Coefficient on the side of the liquid:

$$\delta_F = \left[\frac{10^{-6}}{10^6 \times 9.81} \right]^{1/3} = 0.047.10^{-3} \text{ m}$$

$$LM_L = \frac{1.95}{0.785 \times 0.84^2} = 3.52 \text{ kg.m}^{-2}.s^{-1}$$

$$Re_L = \frac{4 \times 3.52}{10^{-3} \times 28.19} = 500$$

$$Sc_L = \frac{10^{-3}}{10^3 \times 0.20.10^{-8}} = 500$$

$$\beta_L = \frac{55 \times 0.2.10^{-8}}{0.047.10^{-3}} \times 0.012 \times 500^{0.66} \times 500^{0.33}$$

$$\beta_L = 0.01344 \text{ kmol.m}^{-2}.s^{-1}$$

4) Heights of transfer unit:

$$G = \frac{1.5 \times 1}{32} = 0.0469 \text{ kmol.m}^{-2}.s^{-1}$$

$$H_G = \frac{0.0469}{0.00142 \times 28.19} = 1.17 \text{ m}$$

$$L = \frac{3.52}{18} = 0.1955 \text{ kmol.m}^{-2}.s^{-1}$$

$$H_L = \frac{0.1955}{0.01344 \times 28.19} = 0.51 \text{ m}$$

With the direct calculation method:

$$H_G = \frac{1.33}{28.19} \times 4736^{0.36} \times 1.32^{0.66} \times 1.0093^{-1}$$

$$= 0.04712 \times 21.045 \times 1.20 \times 0.99$$

$$H_G = 1.17 \text{ m}$$

$$H_L = 21 \times 0.047.10^{-3} \times 500^{0.34} \times 500^{0.67}$$

$$= 0.987 \times 10^{-3} \times 8.27 \times 64.31$$

$$H_L = 0.51 \text{ m}$$

$$H_{OG} = 1.17 + \left[\frac{1.29 \times 0.0469}{0.1955} \right] \times 0.51$$

$$H_{OG} = 1.32 \text{ m}$$

3.4. Effects of the distribution of the liquid: dead height

3.4.1. *General*

In order for a column to work properly, the following is necessary:

– the injection of liquid at the top of the column must lead to a uniform distribution of that liquid across the section of the column. For this purpose, there are various devices (feeders) in existence. We shall examine the performances of some of these;

– the wall effect must be combatted. Indeed, the liquid always has a tendency to spread horizontally and, consequently, to accumulate at the wall, and in this situation, the contact between the gas and liquid becomes slight.

3.4.2. *Feeders*

The wetting of the packing *at the top of the column* is performed by a feeder, which could be:

– a downward vertical jet of liquid, coaxial with the column;

– a jet issuing from a sprayer and which is therefore in the shape of a full cone. However, sprayers may require a pressure of 1–3 bars;

– a spray corona, which is a horizontal torus shot through with holes in its lower circle. The diameter of the torus is generally half that of the column. This type of corona consumes a small amount of pressure;

– gutters whose edges are sawtoothed to render the overflow of the liquid uniform. These gutters are arranged in parallel to one another. They are used for columns with a very large section.

It must be understood that if, on a section of the column, there is a dry zone and a zone irrigated by the liquid, the gas will favor the path corresponding to the dry zone, which prevents mutual contact between those two fluids.

An article published in the erstwhile journal *Génie Chimique* (for which it has not been possible to find the exact reference) shows that the ratio between the vertical distance traveled Z_m and the horizontal displacement L_h of the spreading liquid is equal to 15:

$$Z_m = 15L_h$$

Thus, over the height Z_m which contributes to the dead height, the transfer of solute between the gas and liquid is very slight, if not in fact non-existent. The height corresponding to Z_m is neutralized, meaning that it is useless. Table 3.7 gives the value of L_h depending on the type of feeder.

Nature of the feeder	L_h
Axial jet of liquid	Radius of the column
Full conical jet of spray	Zero
Spray corona	Half the radius of the column
Parallel gutters	Half the interval between gutters

Table 3.7. *Horizontal displacement of the liquid*

EXAMPLE 3.4.–

Consider a spray corona whose diameter is 0.2 m for a column whose diameter is 0.4 m.

$$L_h = \left(\frac{0.4}{2}\right) \times \frac{1}{2} = 0.1$$

The un-useable height of packing beneath the corona is:

$$Z_I = 15 \times 0.1 = 1.5 \text{ m}$$

3.4.3. Accumulation of liquid at the wall (wall effect)

As it descends down the column, therefore, the liquid, whose flowrate is maximum on the axis of the column, has a tendency to spread and, hence, move towards the wall to form a film which practically produces no exchange at all with the gaseous phase.

When the liquid running along a vertical wall does not have many available paths to take it away from the wall, it tends to accumulate there. In other words, the wall is not highly reflective.

However, it has been possible to render the wall of a column almost reflective and, thereby, increase the effectiveness of the transfer by one of the following two methods:

1) coating the internal face of the wall with a product that is not wetted by the liquid;

2) particularly, shaping the wall like an accordion (Kirschbaum). Indeed, the liquid flowing over inclined surfaces is projected toward the inside of the column.

Unfortunately, these are laboratory curios and, on the industrial level, vessel walls are not reflective.

Over a height dZ, the flux density reaching the wall is:

$$V_p = \frac{1}{\pi D_c} \frac{dW_p}{dZ} \tag{3.5}$$

W_p: flowrate at the wall: kg/s or kmol/s, or indeed m.s^{-1}

D_c: diameter of the column: m

The flux density of the liquid in an empty bed in the column is:

$$V_L = \frac{W_L - W_P}{\pi D_2^2 / 4}$$ [3.6]

W_L : is the liquid flowrate fed in at the top of the column

The lateral resistance to the spreading of the liquid in the vicinity of the wall is proportional to the friction surface – i.e. to a_e with (see section 3.3.2):

$$a_e = \mu a_T - a_I$$

In addition, V_p is proportional to V_L . Thus, we write:

$$V_p = V_L \times \frac{K}{4a_e}$$

Let us use expressions [3.1] and [3.2]. We find:

$$\frac{dW_p}{W_L - W_p} = \frac{K}{a_e} \frac{dZ}{D_c}$$

We then integrate:

$$W_p = W_L \left[1 - \exp\left(-\frac{K}{a_e} \frac{Z}{D_c} \right) \right]$$

We shall take $K = 0.5$.

Look now for the limiting value of Z/D_e so that the liquid at the wall represents no more than 10% of the liquid fed in. This is expressed by:

$$\exp\left(-\frac{0.5}{a_e} \frac{Z}{D_c} \right) \geq 0.9 \text{ so } \frac{Z}{D_c} \leq 0.21 a_e$$

For example, with Raschig rings of 25 mm, a_e is equal to 38 m^{-1} (see Table 3.5).

Thus:

$$\frac{Z}{D_c} \le 0.21 \times 38 = 8$$

3.4.4. Effect of parietal flux on absorption and stripping

Consider a liquid feed W_L with the content x_0 and which needs to exit the column at the specified content x_s. The parietal flowrate rises from zero to W_p from the top to the bottom of the column. If we accept, for simplicity's sake, that this rise is linear, the mean parietal flowrate will be taken as equal to $W_p/2 = \overline{W}_P$. In addition, we shall suppose that the liquid at the wall does not give rise to exchange.

The solute balance is written:

$$x_1(W_L - \overline{W}_P) + x_0 \overline{W}_P = x_s W_L$$

The content x_1 is that of the non-parietal liquid at the outlet:

$$x_1 = \frac{x_s - px_0}{1-p} \quad \text{with} \quad \frac{\overline{W}_P}{W_L} = p$$

1) Consider gas absorption with $p = 0.05$:

$$x_0 = 0.01$$

$$x_s = 0.1$$

$$x_1 = \frac{0.1 - 0.05 \times 0.01}{1 - 0.05} = 0.105$$

We can see that the performance of the column needs to increase from $x_1 = 0.100$ to $x_1 = 0.105$, which does not cause a significant alteration of the equipment.

2) Consider the stripping of a liquid, still with $p = 0,05$:

$$x_0 = 0.1 \quad x_s = 0.01$$

$$x_1 = \frac{0.01 - 0.05 \times 0.1}{1 - 0.05} = 0.0053$$

Owing to the wall effect, the column must be designed with a specification x_1 as half of x_s, which may cause its height to triple, flowrates permitting. Packed columns need to be used carefully for stripping.

NOTE.– According to Leenaerts [LEE 66], the velocity profile of the liquid in an empty bed in a packed column is not uniform, and resemble a bell curve. The local velocity can be expressed as:

$$V_L = G(r)\overline{V}_L$$

$\overline{V}_L$: mean velocity in the section of the column: $m.s^{-1}$

Thus:

– not too far from the axis: $G(r) > 1$,

– in the vicinity of the wall: $G(r) \ll 1$.

Locally, the flux density for the spreading of the liquid is:

$$V_h = kV_L$$

The coefficient k is constant throughout the entire section of the column.

Finally:

$$V_h / \overline{V}_L = kG(r)$$

However, we have seen that:

– not too far from the axis:

$$V_h / \overline{V}_L = L_h / Z = 1/15 = 0.067$$

– on the periphery (Raschig rings of 25 mm):

$$V_h / \overline{V}_L = V_p / \overline{V}_L = \frac{0.5}{4a_e^*} = 0.0033$$

These results, then, are consistent.

3.4.5. *Remedying the wall effect – recenterers*

We divide the column into segments whose height corresponds to approximately ten times the diameter of the column. Between the segments thus defined, we install recentering distributors, which are conical frusta combined with an overflow corona (see Figure 3.1).

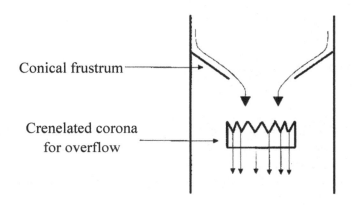

Conical frustrum

Crenelated corona
for overflow

Figure 3.1. *Recenterer with a conical frustum and crenelated corona*

The slope of the conical frustum may be 45° and its internal diameter will be around 0.5 times that of the column. The pressure drop in the whole system is less than 500 Pa. It is due to the changes in direction of the gas. This type of device is recommended for columns whose diameter is less than 0.4m. Indeed, the influence of the wall effect quickly fades when the diameter of the column increases.

3.4.6. *Conclusion*

The results acquired in Chapter 4 concerning material transfers of material are immediately usable for the calculations for a packed column, even in complicated cases such as the adiabatic absorption into water of hydrochloric gas or ammonia. For this purpose, we divide the column into ten or so slices. To calculate the section $k+1$ as a function of the section k, we calculate the material transferred with the composition in k, which gives us an initial value of the composition in $k+1$. We then calculate the matter transferred using that composition. Next, we find the arithmetic mean of these two transferred amounts of material (from k and from k+1), which gives us the value of the composition in k+1 from k. Appendix 1 gives the vapor pressures p_i of HCl, NH_3 and H_2O as a function of the temperature and of the composition of the liquid phase. If P_T is the total pressure, we have:

$$y_i^* = \frac{p_i}{P_T}$$

The Runge–Kutta method (see Appendix 4) represents an improvement for the solving of this problem.

Remember, though, that the old concept of the HETP (height equivalent to a theoretical plate) can be evaluated as equal to around 25 times the nominal size of the packing. Thus, we can quickly gain a rough estimation of the height of a distillation column or even an isothermal absorption column. However, this way of working is imprecise, because there is no rational justification for the concept of the HETP.

Batch Distillation

4.1. Simple boiling

4.1.1. *Simple boiling of a vat*

The ASTM procedure of petroleum engineers is a standardized version of this laboratory operation. The equipment necessary for this procedure includes only a balloon with a thermometer, heated electrically. As a function of the fraction vaporized, we raise the temperature of the liquid in the balloon, and therefore in the vat. This way of working, without a column above the vat, is similar to the distillation in an alembic or the retorts used by alchemists.

Such an operation can be performed by heating a simple balloon and noting the temperature of the liquid as it vaporizes. At each time τ, the liquid is at its boiling point, and an equilibrium calculation gives us the instantaneous composition y_i of the vapor produced.

Over the course of a time period $\Delta\tau$, the molar quantity ΔV of vapor produced is such that:

$$Q\Delta\tau = \Delta V \sum_i y_i \left(H_i - h_i \right)$$

Q : thermal power of heating of the vat: Watt

H_i and h_i : enthalpies of the component i in the vapor state and in the liquid state: $J.kmol^{-1}$

The amount of component i present in the vat is $m_i(\tau)$ and over the period $\Delta\tau$, this quantity decreases by:

$$\Delta m_i = \Delta V y_i$$

The composition of the vat has become:

$$x_i = \frac{m_i - \Delta m_i}{\sum_i (m_i - \Delta m_i)}$$

Using this composition, at the boiling point of the vat, we are able to calculate the y_i at equilibrium with the x_i. Let (x) symbolize the set of x_i values. The above relations are of the form:

$$\frac{d(x)}{d\tau} = F\ ((x), \tau)$$

The Runge–Kutta procedure can be employed (see Appendix 4).

NOTE (Purification of dirty liquids).–

The purification of used motor oils can be done by simple boiling in a moderate vacuum, followed by condensation of the oil vapors. Indeed, the ultra-fine particles of carbon which blacken the oil have a vapor pressure of zero. The boiler must be composed of flat parallel hollow plates, containing electrical resistors, or else, more economically, allowing the passage of water vapor heating. Two or three internal chicanes ensure the rigidity of the hollow plate and distribute the vapor evenly. The flatness of the plates and sufficient spacing between them make for easy cleaning, because the boiling of a dirty liquid inevitably leaves deposits on the heated surfaces.

Purification by filtration is not a viable solution, for the simple reason of quick clogging of the filtering support.

The most elegant solution is centrifugal decantation, with a machine made up of superposed "plates", rotating around a vertical axis which passes through their center. However, these machines are not cheap to buy, and this method does not eliminate certain dissolved products deriving from the degradation of additives in the oil, whereas simple boiling can probably do the job well.

In any case, the oil thus recovered can only, at best, be used as a filler in the manufacture of new oil, unless we obtain a detailed analysis of the components of the recovered oil and define an appropriate range of adjuvants to add to it in order to make it usable.

4.2. Total-reflux distillation

4.2.1. *Operating principle*

In steady operation, the components rotate in the column, rising with the vapor and dropping with the liquid. The volatile products are primarily drawn *toward the top of the column* (high values of x_{ji} and y_{ji} for these products), whereas the heavier fractions are found *at the bottom* (high values of x_{ji} and y_{ji} for these heavy products and small values for the light ones).

At the bottom of the loop corresponding to each component, that component is transferred from the liquid to the vapor phase, and at the top, the transfer takes place from the vapor to the liquid.

4.2.2. *Important convention*

Throughout this volume just like everywhere else in this set of books, the plates will be numbered from top to bottom of the column with the index $j = 1$ for the ensemble of the condenser and the reflux tube, and the index $j = p$ for the vat and its boiler. Additionally, Appendix 5 discusses the implementation of the total molar quantities M_j on the plates.

4.2.3. *Balances and transfers of material and heat (total reflux)*

Consider a domain encapsulating the vat and its boiler, as well as the bottom of the column situated beneath plate $j - 1$. By writing that what goes in must come out (i.e. that the input and output are equal to one another), we obtain, for the component i:

$$\ell_{j-1,i} = v_{j,i} \quad 2 \leq j \leq p$$

The overall flowrates are the sum of the partial flowrates for all the components:

$$L_{j-1} = \sum_i l_{j-1,i} \text{ and } V_{j,i} = \sum_i v_{j,i}$$

Thus:

$$L_{j-1} = V_j$$

It is important to quash the idea that, with total reflux, there is no exchange between the two phases. Indeed:

$$l_{j-1,i} = v_{j,i}$$

$$l_{j,i} = v_{j+1,i}$$

By subtracting, term by term:

$$l_{j,i} - l_{j-1,i} = v_{j+1,i} - v_{j,i} = \Delta w_{j,i}$$

The term $\Delta w_{j,i}$ is the molar quantity of component i (gained by the liquid and lost by the vapor) on plate j, per unit time.

The same reasoning applies to heat:

$$Q_p + L_{j-1} h_{j-1} = V_j H_j$$

(with $Q_p + Q_1 = 0$)

$$Q_p + L_j h_j = V_{j+1} H_{j+1}$$

Thus, by subtraction, we find:

$$L_j h_j - L_{j-1} h_{j-1} = V_{j+1} H_{j+1} - V_j H_j = \Delta q_j$$

The term Δq_j is, on plate j, the thermal power gained by the liquid and lost by the vapor.

On each plate, we need to determine:

– the material exchanged between the liquid and vapor,

– the variations in temperature of the liquid and the vapor.

To do so, we refer to the discussion in Chapter 2 for real plates and to Chapter 3 for packed columns.

4.2.4. Calculations for a total reflux column

This calculation must be performed in several iterations:

1) calculation from the vat, supposing that the molar quantity present on the plates is null. However, from this, we deduce the compositions of the liquid on the plates;

2) taking molar quantities of liquid on the plates and, using the calculated compositions, we can, by subtraction, discover the molar quantity and composition of the vat;

3) we repeat the calculation for the column, starting with the vat, which gives new compositions of the liquid on the plates;

4) we correct the molar quantity and composition of the vat;

and so on.

We shall proceed on the basis of the vat kept at boiling point. An equilibrium calculation gives us the composition and temperature of the vapor V_p. The composition of the liquid L_{p-1} is the same as that of V_p. On the other hand, it is necessary to make a hypothesis about the temperature of that liquid which can be estimated as 2 or 3°C below the boiling point of that liquid. Let $h_{p-1}^{(0)}$ represent the corresponding enthalpy.

The calculation can then be easily performed, working back up the column until we reach plate 2 which, itself, defines the vapor V_2 and its temperature.

As is common, the condenser will be supposed to operate by total condensation, so that the supercooling of the condensate L_1 will be either zero or imposed, which is tantamount to defining the enthalpy $h_1^{(0)}$ of that condensate. Because we have assimilated h_{p-1} to the value $h_{p-1}^{(0)}$, it is normal that we have:

$$h_1^{(0)} \neq h_1$$

The true value of h_{p-1} must be such that we have:

$$Q_p = V_p H_p - L_{p-1} h_{p-1} = V_2 H_2 - L_1 h_1 = -Q_1$$

However:

$$L_{p-1} = V_p \text{ and } L_1 = V_2$$

Thus:

$$Q_p = V_p \left(H_p - h_{p-1}^{(1)} \right) = V_2 \left(H_2 - h_1^{(0)} \right)$$

This relation enables us to calculate $h_{p-1}^{(1)}$, which should replace $h_{p-1}^{(0)}$. However, it would undoubtedly be prudent to adopt:

$$\overline{h_{p-1}^{(1)}} = \frac{1}{2} \left(h_{p-1}^{(1)} + h_{p-1}^{(0)} \right)$$

In any case, three bottom-up calculations for the column will certainly be sufficient.

4.3. Batch distillation

4.3.1. Definitions and advantage to the operation

In order for continuous distillation to become discontinuous, we simply need to suddenly change some of the parameters. For this reason, here, "discontinuous distillation" will be different from "batch distillation". The equipment we shall study will contain:

– a vat heated electrically or with a steam boiler;

– a plate column or packed column above the vat;

– a condenser cooled by refrigerant liquid, with a reflux tube to collect the condensate.

This equipment contains no external feed or withdrawal mechanism, other than the withdrawal of distillate at the outlet from the reflux tube. In principle, distillation is continued until the vat is exhausted – i.e. until all the liquid contained in the vat has gone. The study presented below is the theory of the true boiling point (TBP) curve used by American Petroleum Engineers. This curve represents the variations in the temperature of the distillate in the vapor state at the top of the column as a function of the fraction of the vat's contents having been vaporized. If the column is equivalent to 30 theoretical plates and if the reflux rate is around 30 (i.e. very high), the TBP appears as a succession of platforms, each corresponding to a pure component.

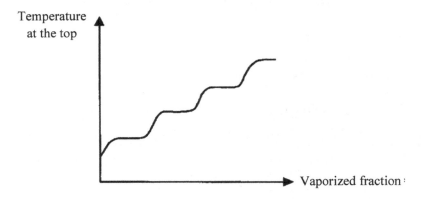

Figure 4.1. *Shape of the TBP curve for a mixture of petroleum spirits*

Thus, the components exit successively with the distillate, a little like the effluents in chromatography.

Now suppose that we have the following conditions:

– the amount being handled is small;

– the number of components is limited;

– low purity is sufficient for the products obtained (a component of which there is little in the vat would be obtained with mediocre purity).

The fractioning will be all the better when there is a broad range of fugacities. We could, for instance, design a $5\,m^3$ vat, topped with a recovery column 0.8 m in diameter and containing 20 or 30 plates. It may therefore be cheaper to distill several vats over the course of a year rather than install three or four columns running continuously.

The provisional calculation method proposed here is an improvement of that published by Domenech and Enjalbert [DOM 81]. Indeed, here, the plates are real rather than theoretical, so the mean composition of the liquid on a plate is no longer equal to the output composition.

4.3.2. Time taken to cross a plate

Let us set down the following definitions:

S_p : spacing between two consecutive plates: m

h_{LC} : height of clear liquid on the plate: m

V_G : velocity of the gas in an empty bed: $m.s^{-1}$

V_L : velocity of the liquid in an empty bed: $m.s^{-1}$

The order of magnitude of the crossing times is:

$$\tau_L \# \frac{h_{LC}}{V_L} = \frac{0,02}{10^{-3}} = 20\,s \text{ and } \tau_G \# \frac{S_p - h_{LC}}{V_G} = \frac{0,4 - 0,02}{1} = 0,38\,s$$

Thus, any disturbance in the gaseous flowrate immediately impacts across the whole height of the column. On the other hand, a disturbance of the liquid flowrate at the top will take several minutes to have an impact at the bottom of the column. Similar conclusions can be drawn for packed columns.

4.3.3. Choice of time increment

The increment $\Delta\tau_k$ separates times τ_{k-1} and τ_k .

Let us set:

M_p^0 : initial molar quantity in the vat

D : molar flowrate of the distillate withdrawn

The time T of the rectification is less than the following maximum:

$$T < T_{max} = M_p^0 / D$$

Indeed, when the liquid level in the vat is relatively low, the boiler begins not to work properly, because the remaining heavy products are often difficult to fractionate.

If n is the predictable number of components present in the mixture being processed, we set:

$$\Delta \tau_k = \frac{T}{qn}$$

The number q is greater than or equal to 3 or 4, in the hope of clearly seeing the appearance of platforms of temperature and composition at the top of the column.

4.3.4. Balances of the vat + boiler ensemble

Suppose we know the overall flowrate, the composition and the enthalpy of the liquid exiting plate $p-1$ at time $k-1$, and the composition and overall molar quantity of the vat's contents.

The material balances of the vat are written:

– overall balance:

$$\frac{M_p^k - M_p^{k-1}}{\Delta \tau_k} = L_{p-1}^{k-1} - V_p^k$$

The overall flowrate of vapor will be:

$$V_p^k = L_{p-1}^{k-1} + D$$

Indeed, the overall molar quantities M_j are constant in the reflux tube and throughout the column;

– partial balances:

$$\frac{m_{p,i}^k - m_{p,i}^{k-1}}{\Delta\tau} = L_{p-1}^{k-1}x_{p-1,i}^{k-1} - V_p^k y_{p,i}^k$$

$$M_p = \sum_i m_{p,i} \quad x_{p,i} = \frac{m_{p,i}}{M_p}$$

The composition of the vapor V_p results from the equilibrium between that vapor and the vat contents. The composition of the mixture would be the arithmetic mean of $x_{p,i}$ at time $k-1$ and at time k. The result of this is that the equilibrium calculation is iterative. Note that this calculation gives us the temperature and enthalpy H_p of the vapor V_p, and similarly for h_p, the enthalpy of the content of the vat all at time k.

The heat balance of the vat + boiler ensemble is written:

$$Q_p + L_{p-1}h_{p-1} = V_p H_p + \frac{d(M_p h_p)}{d\tau}$$

However:

$$\frac{d(M_p h_p)}{d\tau} = h_p \frac{dM_p}{d\tau} + M_p \frac{dh_p}{d\tau} \text{ and } \frac{dM_p}{d\tau} = -D$$

The balance becomes:

$$Q_p = V_p^k \left(H_p^k - h_{p-1}^{k-1} \right) - D\left(h_p^k - h_{p-1}^{k-1} \right) + M_p \frac{h_p^k - h_p^{k-1}}{\Delta\tau}$$

From this balance comes the thermal power Q_p of the boiler.

4.3.5. *Balances of the condenser + reflux tube ensemble*

Suppose, for the vapor V_2, we know its overall flowrate, composition and enthalpy (or temperature). In addition, the liquid in the reflux tube is homogenous.

The material balance for the reflux tube is:

$$L_1^+ = V_2 \quad L_1^+ = L_1 + D$$

$$L_1^+ x_{1,i} = V_2 y_{2,i} - M_1 \frac{dx_{1,i}}{d\tau}$$

Between times k–1 and k, we have:

$$\frac{1}{2}\left(x_{1,i}^k + x_{1,i}^{k-1}\right) L_1^{+k} = V_2^k y_{2,i}^k - M_1 \frac{\left(x_{1,i}^k - x_{1,i}^{k-1}\right)}{\Delta\tau_k} \quad \text{where } L_1^{+k} = V_2^k$$

From the above equation, we deduce that: $x_{1,i}^k$.

The heat balance for the reflux tube is written:

$$\frac{1}{2}\left(h_1^k + h_1^{k-1}\right) L_1^{+k} = L_1^{+k} h_1^{ck} - M_1 \frac{\left(h_1^k - h_1^{k-1}\right)}{\Delta\tau_k}, \text{ from which we obtain } h_1^k$$

The enthalpy h_1^{ck} is that of the condensate exiting the condenser. That condensate is at its boiling point.

The downward reflux in the column is:

$$L_1 = L_1^+ - D = \frac{V_2 R}{R+1} \quad \text{(R is the reflux rate)}$$

Note that the instantaneous refrigeration power of the condenser is given by:

$$Q_1^k = V_2^k \left(H_2^k - h_1^{ck}\right)$$

The heat balance for the reflux tube is:

$$M_1 \frac{\left(h_1^k - h_1^{k-1}\right)}{\Delta\tau_k} = L_1^+ \left(h_1^{ck} - h_1^k\right)$$

Hence, we have the enthalpy h_1^k of the liquid exiting the reflux tube and entering the column.

4.3.6. *Proceeding of the simulation*

Initial conditions: the column works with total reflux.

Time τ_1: between the start of the withdrawal of the distillate and the arrival of the disturbance in the liquid flowrate in the vat, a period of time elapses of around 1 to several minutes. During that time, we can accept that the composition and temperature of the distillate are those of the condensate during total reflux.

Time τ_2: the mixture begins to be depleted. The balances for the vat and the boiler give the vapor V_p at time τ_1 and, based on the liquid known L_{p-1} at time τ_0, we work back up the column, using the transfers and balances of heat and material for each plate. The material and heat balances of the condenser + reflux tube ensemble give the liquid L_1 at time τ_1, in terms of its composition, flowrate and temperature. We work back down the column to the vat, which gives us a new value of L_{p-1} at time τ_1 and, having found the balances for the vat, we work back up to the condenser. Two or three journeys up and down the column suffice for convergence to be achieved, so that the characteristics of L_1 and L_{p-1} no longer change.

It may be of interest to approach the solution more slowly, by choosing not the new values of L_{p-1} and L_1, but their arithmetic mean with the previous value.

Time τ_3, τ_4...: the procedure is the same as at time τ_2.

In summary, for each time period $\Delta\tau_k = \tau_k - \tau_{k-1}$, we perform:

1) the calculation of the balances in the vat;

2) the climbing of the column;

3) the calculation of the balances of condensate;

4) the descent of the column and return to 1 until we have convergence for L_1 and L_{p-1}.

Appendix 5 describes the way to determine the molar hold-up of liquid on the plates. The residence time of the liquid in the reflux tube is around 5 minutes.

4.3.7. Practical running of an industrial installation

Over time, the vat and the column become poorer in light components and, consequently, richer in the heavier components. Thus, the boiling and dew points of the liquid and gaseous mixtures present gradually increase, which has consequences for the operation of the boiler and the condenser. It is desirable, though, throughout the operation, to be able to adjust the thermal powers Q_1 and Q_p of the condenser and boiler at will.

In a condenser, the heat transferred from the vapor to the coolant liquid increases with the gap in the temperatures of the two fluids. If the proportion of heavy components increases, the dew and boiling points of the vapor and its condensate increase, which increases the temperature gap. The solution is to decrease the flowrate of coolant liquid (line (2) in Figure 4.2).

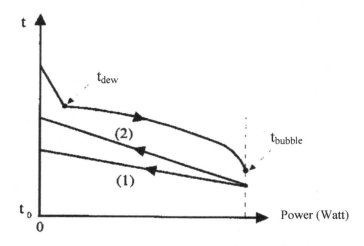

Figure 4.2. *Heat transfer curves for total condensation*

The boiler is composed of a bundle of vertical pipes, generally 2 m high and with an internal diameter of 2 cm. The heating vapor condenses on the outside of the pipes, and it is in those tubes that the boiling takes place. From the very start of the operation, the device must be overdimensioned. If the lower part of the tubes is flooded with the condensate of the heating vapor, the boiler, whilst it may be oversized, will deliver the correct amount of thermal power. We merely need to gradually lower the level of condensate to increase the device's power. Naturally, if the heating system is electrical, we simply need to set the position of a rheostat to obtain the desired power.

Thus, it is possible to obtain preset values of Q_1 and Q_p.

Researchers in laboratories and industrial operators seek to obtain the following during the course of the operation:

– a constant distillate flowrate D, which can easily be obtained by using a flow-regulating valve;

– a constant reflux rate L_1/D of around 20–30. For this purpose, we need to act on the thermal powers Q_1 and Q_p, as has just been explained.

APPENDICES

Appendix 1

A Few Expressions for Partial Vapor Pressures

A1.1. Henry's constant

Henry's law is expressed by:

$$p = Hx$$

p: vapor pressure of the solvent at equilibrium

x: molar fraction of the solute

p is measured in atmospheres,

H is therefore measured in atmospheres, and depends on the temperature. Thus, for ammonia dissolved in water at a concentration of less than 15%:

$$H = \exp\left(12.186 - \frac{3031.16}{t + 229.35}\right)$$

H: atm

$$t = {}^{\circ}C$$

When using Henry's law, it is important to remember that, in parallel, the vapor pressure of the solvent varies in accordance with Raoult's law, meaning that it is proportional to the molar fraction of the solvent.

A1.2. Empirical formulae

In certain cases – e.g. when there is complexation in solution or even a chemical reaction – it is necessary to employ empirical expressions.

Such is the case with the dissolution of ammonia gas and hydrochloric gas in water (hereinafter, the x values represent the mass fractions).

System NH_3 – water : contents less than 15%: use Henry's law.

NH_3 : contents greater than or equal to 15%, use the following relation:

$$p = \exp\left[\left(11.01672 - \frac{2814.219}{t+230}\right)\right] \bigg/ \left(\frac{1-1.6x}{x}\right)\left(0.7574 + \frac{62.51}{t+230}\right)$$

Water:

$$p = \exp\left[13.778154 + 0.267373x - 2.74972x^2\right.$$

$$\left. - \left(\frac{5185.22 - 52.3998x + 156.448x^2}{t+273}\right)\right]$$

HCl-water system

HCl:

$$p = \exp\left[20.24525 - 14.31977x - \left(\frac{11121.269 - 22170.193x + 17972.592x^2}{t+273}\right)\right]$$

Water:

$$p = \exp\left[13.88267 - 1.0834x - \left(\frac{5326.137 - 21262.392x + 7290.398x^2}{t+273}\right)\right]$$

For both systems:

p is expressed in atmospheres,

x is expressed in *gravimetric* fractions.

Appendix 2

Characteristics of Typical Packings

Material	Dimension (mm)	Thickness of the wall (mm)	Apparent density $(kg.m^{-3})$	Volumetric surface (m^{-1})	No-load fraction (%) (Porosity)
Ceramic Raschig rings	102	11	600	50	75
	76	9.5	650	70	75
	51	6.5	650	95	75
	38	6.5	700	130	68
	25	3	700	200	73
	19	2.5	700	240	72
	13	2.5	800	370	64
	9.5	1.5	800	500	65
	8	1	800	600	72
	6.5	0.8	800	800	70
Steel Raschig rings	76	1.6	450	70	94
	51	1.2	460	100	94
	38	0.9	480	140	94
	25	0.7	560	210	93
	19	0.6	580	270	93
	13	0.5	700	400	91
	9.5	0.5	930	600	88
	6.3	0.5	1 400	800	82

Ceramic Pall rings	102	9.5	420	56	82
	51	5	550	125	78
	25	3	640	220	73
Steel Pall rings	51	1	400	105	95
	35	0.8	430	145	95
	25	0.6	500	240	94
	16	0.4	550	370	93
Porcelain Berl saddles	51		640	110	77
	38		610	150	76
	25		720	250	70
	19		800	300	67
	13		900	480	65
	6.3		900	1 000	62
Intalox saddles	51		600	110	75
	38		600	160	74
	25		600	250	75
	19		600	300	73
	13		600	480	73

Appendix 3

Proposal For a Normalized Foaming Test

A3.1. Experimental protocol

Take four clean test tubes (washed with distilled water).

Fill the tubes to one third of their height with the liquid to be tested and four control solutions. Stir the tubes for 5 seconds, parallel to their axes. By comparing the test liquid with the control solutions, we are able to assign the liquid a foaming index.

A3.2. Composition of the control solutions

Foaming index	Composition of the solution
4 Highly foamy	2 drops of Teepol in water in 1/3 of test tube
3 Foamy	1 drop of Teepol in water in 1/3 of test tube
2 Slightly foamy	Solution 3 is diluted with an equal volume of water and expressed in relation to 1/3 of the tube
1 Non-foamy	Water with no Teepol

Appendix 4

Numerical Integration: Runge–Kutta 4th Order Method

Integrate the following differential equation:

$$\frac{dx}{d\tau} = F(x, \tau)$$

$$x_{\tau=0} = x_0$$

We assume:

$$\tau_{i+1/2} = \tau_i + \frac{\Delta\tau}{2}$$

$$x_{i+1/2}^{(1)} = x_i + \frac{\Delta\tau}{2} F(x_i, \tau_i)$$

$$x_{i+1/2}^{(2)} = x_i + \frac{\Delta\tau}{2} F(x_{i+1/2}^{(1)}, \tau_{i+1/2})$$

$$x_{i+1}^{(1)} = x_i + \Delta\tau\, F(x_{i+1/2}^{(2)}, \tau_{i+1/2})$$

where:

$$x_{i+1} = x_i + \frac{\Delta\tau}{6}\Big[F(x_i, \tau_i) + 2F(x_{i+1/2}^{(1)}, \tau_{i+1/2}) +$$

$$2F(x_{i+1/2}^{(2)}, \tau_{i+1/2}) + F(x_{i+1}^{(1)}, \tau_{i+1})\Big]$$

We can apply a general method to a system of n differential equations of the first-order, involving n variables x_j (j from 1 to n). The independent variable is x_0:

$$\frac{dx_j}{dx_0} = F_j\left(x_0, x_1, ..., x_j, ..., x_n\right)$$

Assume:

$$x_{0,i+1} = x_{0,i} + \Delta x_0 \qquad \text{and} \quad x_{0,i+1/2} = x_{0,i} + \frac{\Delta x_0}{2}$$

$$x^{(1)}_{j,i+1/2} = x_{j,i} + \frac{\Delta x_0}{2} F_j\left(x_{0,1}, ..., x_{j,i}, ...x_{n,i}\right)$$

$$x^{(2)}_{j,i+1/2} = x_{j,i} + \frac{\Delta x_0}{2} F_j\left(x_{0,i+1/2}, ..., x^{(1)}_{j,i+1/2}, ..., x^{(1)}_{n,i+1/2}\right)$$

$$x^{(1)}_{j,i+1} = x_{j,i} + \Delta x_0 F\left(x_{0,i+1/2}, ..., x^{(2)}_{j,i+1/2}, ..., x^{(2)}_{n,i+1/2}\right)$$

And, finally:

$$x_{j,i+1} = x_{j,i} + \frac{\Delta x_0}{6}\left[F_j\left(x_{0,i}, ...x_{j,i}, ..., x_{n,i}\right) + 2F_j\left(x_{0,i+1/2}, ...x^{(1)}_{j,1+1/2}, ...x^{(1)}_{n,i+1/2}\right) + \right.$$
$$\left. 2F_j\left(x_{0,i+1/2}, ...x^{(2)}_{j,i+1/2}, ...x^{(2)}_{n,i+1/2}\right) + F_j\left(x_{0,i+1}, ...x^{(1)}_{j,i+1}, ...x^{(1)}_{n,i+1}\right)\right]$$

Appendix 5

Molar Retentions

In each area, the liquid volume Ω flows according to the existing hydrodynamic conditions in the area. If $\overline{\omega}_{ji}$ is the partial volume of component i in the liquid of tray j, we should find, at time k:

$$\sum_i \overline{\omega}_{ji}^k m_{ji}^k = \Omega_j = \text{cste}$$

where m_{ji}^k represents the molar quantity of component i present on the tray j at time k.

Therefore, after this calculation, we generally find:

$$\sum_i \overline{\omega}_{ji}^k m_{ji}^{k(0)} = \Omega_j^{(0)} \neq \Omega_j$$

According to which we must revise $m_{ji}^{k(0)}$ and insert:

$$m_{ji}^k = m_{ji}^{k(0)} \frac{\Omega_j}{\Omega_j^{(0)}}$$

The molar quantity present in area j at time k is therefore:

$$M_j^k = \sum_i m_{ji}^k$$

However, the molar quantity present in the tank gradually decreases as the distillation continues and there is no correction make it at this level.

These calculations can be broken down into the following steps:

– calculate the total reflux using the assumed retentions $M_j^{(0)}$;

– calculate the total concentration $c_{T,j}^{(0)}$ (kmol.m^{-3}) throughout the column;

– use the hydrodynamics of the column to calculate the volumetric retentions Ω_j (j from 1 to p–1; where p is the tank)

– recalculate the molar retentions by $M_j^{(1)} = \Omega_j c_{T,j}^{(0)}$

$M_j^{(1)}$ retentions are assumed to be constant for every calculation of tank distillation.

Bibliography

[BEN 83] BENETT D.L., AGRAWAL R., COOK P.J., "New pressure drop correlation for sieve tray distillation columns", *American Institute of Chemical Engineers*, vol. 3, p. 434, 1983.

[BOL 56] BOLLES W.L., "Optimum bubble-cap tray design", *Petroleum Processing*, vol. 11, nos. 2–5, 1956.

[DAN 55] D'ANCONA H.C., HANSON D.N., WILKE C.R., "Capacity factors in the performance of perforated plate columns", *American Institute of Chemical Engineers*, vol. 4, p. 441, 1955.

[DOM 81] DOMENECH S., ENJALBERT M., "Program for simulating batch rectification as a unit operation", *Computers and Chemical Engineering*, vol. 5, no. 3, p. 181, 1981.

[DUR 16] DUROUDIER J.P., *Thermodynamics*, ISTE Press, London and Elsevier, Oxford, 2016.

[ECK 58] ECKERT J.S., FOOT E.H., HUNTINGTON R.L., "Pall rings – new type of towers", *Chem. Eng. Progress*, vol. 54, no. 1, p. 70, 1958.

[ECO 78] ECONOMOPOULOS A.P., "Computer design of sieve trays and tray columns", *Chemical Engineering*, vol. 85, no. 27, pp. 109–120, 1978.

[EDM 61] EDMISTER W.C., *Applied Hydrocarbon Thermodynamics*, Gulf Publishing Company, 1961.

[FAI 61] FAIR J.R., "How to predict sieve tray entrainment and flooding", *Petro. Chem. Engineer.*, vol. 33, no. 10, p. 45, 1961.

[FAI 63] FAIR J.R., "Tray hydraulics: erforated trays", in Smith B.D., *Design of Equilibrium Stage Processes*, McGraw Hill, New York, 1963.

[FEL 82] FELL C.J.D., PINCZEWSKI W.V., "Coping with entrainment problems in low and moderate pressure distillation columns", *Instit. Chem. Engrs. Symp. Series*, no. 73, p. D1, 1982.

[HOF 64] HOFTYZER P.J., "Liquid distribution in a column with dumped packing", *Trans. Inst. Chem. Engrs*, vol. 42, p. T109, 1964.

[HUG 57] HUGHMARK G.A., O'CONNELL H.E., "Design of perforated plate fractionating towers", *Chem. Eng. Progr.*, vol. 53, no. 3, p. 127, 1957.

[JER 73] JERONIMO M.A., SAWISTOWSKI H., "Phase inversion correlation for sieve trays", *Trans. Inst. Chem. Engrs*, vol. 51, p. 265, 1973.

[JES 43] JESSER B.W., ELGIN J.C., "Studies of liquid holdup in packed towers", *Trans. Am. Inst. Chem. Eng.*, vol. 39, p. 277, 1943.

[LEE 66] LEENAERTS R., "Le profil d'écoulement de la phase liquide dans les colonnes à garnissage", *Génie Chimique*, vol. 96, no. 4, p. 947, 1966.

[LEI 57] LEIBSON I., KELLEY R.E., BULLINGTON L.A., "How to design perforated trays", *Petroleum Refiner*, p. 127, February 1957.

[LOC 84] LOCKETT M.J., BANIKS., "Weeping from sieve trays", *AIChE Anuual Meeting*, San Francisco, November 1984.

[LOC 86] LOCKETT M.J., *Distillation Tray Fundamentals*, Cambridge University Press, 1986.

[LOO 73] LOON R.E., PINCZEWSKI W.V., FELL C.J.D., "Dependence of the froth to spray transition on sieve tray design parameters", *Trans. Inst. Chem. Engrs*, vol. 51, p. 374, 1973.

[MCA 63] MCADAMS W.H., *Heat Transmission*, McGraw Hill, 1963.

[MUL 57] MULLIN J.W., "The effect of maldistribution on the performance of packed columns", *The Industrial Chemist*, p. 408, August 1957.

[NOU 85] NOUGIER J.P., *Méthodes de calcul numérique*, Editions Masson, 1985.

[PAR 58] PARIS, *Les procédés de rectification dans i'industrie chimique*, Editions Dunod, 1958.

[PRA 69] PRAHL W.L., "Pressure drop in packed columns", *Chemical Engineering*, p. 89, 11 August 1969.

[SIG 75a] SIGALES B., "How to design reflux drums", *Chemical Engineering*, pp. 157–160, 3 March 1975.

[SIG 75b] SIGALES B., "More on design of reflux drums", *Chemical Engineering*, pp. 87–90, 29 September 1975.

[STI 78a] STICHLMAIR J., MERSMANN A., "Dimensioning plate columns for absorption and rectification", *Inst. Chem. Eng.*, vol. 18, no. 2, p. 223, 1978.

[STI 78b] STICHLMAIR J., *Grundlagen der dimensionierung des gas/flussighkeit kontaktaparates bodenkolonne*, Wiley-VCH, 1978.

[STI 98] STICHLMAIR J.G., FAIR J.R., *Distillation: Principles and Practice*, Wiley-VCH, 1998.

[TAO 63] TAO L.C., "Equation sets size of packed column", *Hydroc. Process. Petroleum Refiner*, vol. 42, no. 9, p. 205, 1963.

[TAY 69] TAYLOR D.L., EDMISTER W.C., "General solution for multi-component distillation processes", *Proceedings of the International Symposium on Distillation*, Brighton, UK, 1969.

[TRE 68] TREYBAL R.E., *Mass Transfer Operations*, 1st ed., McGraw Hill, 1968.

[TRE 80] TREYBAL R.E., *Mass transfer operations*, 2nd ed., McGraw Hill, 1980.

[VAN 47] VAN KREVELEN D.W., HOFTIJZER P.J., "Studies of gas absorption", *Recueil des Travaux Chimiques des Pays-Bas*, vol. 66, p. 49, 1947.

[WUI 65] WUITHIER P., *Raffinage et génie chimique*, Editions Technip, 1965.

[WUI 72] WUITHIER P., *Raffinage et génie chimique*, 2nd ed., Editions Technip, 1972.

[YOS 58] YOSHIDA F., KOYANAGI T., "Liquid phase mass transfer rates and effective interfacial area in packed absorption columns", *Ind. and Eng. Chemistry*, vol. 50, no. 3, p. 365, 1958.

[ZEN 54] ZENZ F.A., "Calculate capacities of perforated plates", *Petroleum Refiner*, pp. 99–102, February 1954.

[ZUI 82] ZUIDERWEG F.J., "Sieve Trays. A view on the state of the art", *Chem. Eng. Science*, vol. 37, no. 10, p. 1441, 1982.

Index

Printed in the United States
By Bookmasters